Practical Splunk Search Processing Language

A Guide for Mastering SPL Commands for Maximum Efficiency and Outcome

Karun Subramanian

apress®

Practical Splunk Search Processing Language: A Guide for Mastering SPL Commands for Maximum Efficiency and Outcome

Karun Subramanian
Greater Minneapolis, MN, USA

ISBN-13 (pbk): 978-1-4842-6275-7				ISBN-13 (electronic): 978-1-4842-6276-4
https://doi.org/10.1007/978-1-4842-6276-4

Copyright © 2020 by Karun Subramanian

This work is subject to copyright. All rights are reserved by the Publisher, whether the whole or part of the material is concerned, specifically the rights of translation, reprinting, reuse of illustrations, recitation, broadcasting, reproduction on microfilms or in any other physical way, and transmission or information storage and retrieval, electronic adaptation, computer software, or by similar or dissimilar methodology now known or hereafter developed.

Trademarked names, logos, and images may appear in this book. Rather than use a trademark symbol with every occurrence of a trademarked name, logo, or image we use the names, logos, and images only in an editorial fashion and to the benefit of the trademark owner, with no intention of infringement of the trademark.

The use in this publication of trade names, trademarks, service marks, and similar terms, even if they are not identified as such, is not to be taken as an expression of opinion as to whether or not they are subject to proprietary rights.

While the advice and information in this book are believed to be true and accurate at the date of publication, neither the authors nor the editors nor the publisher can accept any legal responsibility for any errors or omissions that may be made. The publisher makes no warranty, express or implied, with respect to the material contained herein.

Managing Director, Apress Media LLC: Welmoed Spahr
Acquisitions Editor: Joan Murray
Development Editor: Laura Berendson
Coordinating Editor: Jill Balzano

Cover image designed by Freepik (www.freepik.com)

Distributed to the book trade worldwide by Springer Science+Business Media LLC, 1 New York Plaza, Suite 4600, New York, NY 10004. Phone 1-800-SPRINGER, fax (201) 348-4505, e-mail orders-ny@springer-sbm.com, or visit www.springeronline.com. Apress Media, LLC is a California LLC and the sole member (owner) is Springer Science + Business Media Finance Inc (SSBM Finance Inc). SSBM Finance Inc is a **Delaware** corporation.

For information on translations, please e-mail booktranslations@springernature.com; for reprint, paperback, or audio rights, please e-mail bookpermissions@springernature.com.

Apress titles may be purchased in bulk for academic, corporate, or promotional use. eBook versions and licenses are also available for most titles. For more information, reference our Print and eBook Bulk Sales web page at http://www.apress.com/bulk-sales.

Any source code or other supplementary material referenced by the author in this book is available to readers on GitHub via the book's product page, located at www.apress.com/9781484262757. For more detailed information, please visit http://www.apress.com/source-code.

Printed on acid-free paper

Dedicated to my wife Vani and my kids Sarvajith, Samy, and Nemo for filling my life with love

Table of Contents

About the Author xiii

About the Technical Reviewer xv

Acknowledgments xvii

Introduction xix

Chapter 1: Introducing the Splunk Platform 1

    Machine Data 1

        What Is Machine Data? 2

        Events 3

        Logs 3

        Traces 4

        Metrics 5

        Time-Series Nature of Machine Data 5

        The Value of Machine Data 5

    The Shortcomings of Machine Data 7

        Size 7

        Speed 7

        Structure 8

        Distribution 8

    The Splunk Operational Data Intelligence Platform 8

        Primary Functions of Splunk 8

        Architecture of Splunk Platform 10

    Introducing Splunk Search Processing Language (SPL) 12

        Syntax of SPL 12

        The Search Pipeline 17

TABLE OF CONTENTS

 Navigating the Splunk User Interface 18
  Installing Splunk 19
  Logging onto Splunk Web 19
 Write Your First SPL Query 23
  Using Splunk Tutorial Data 23
  Turning on Search Assistant 28
  Search Modes 30
  Run the Search 31
 Key Takeaways 37

Chapter 2: Calculating Statistics 39
 Stats 39
  A Quick Example 40
  Syntax 40
  Counting Events 42
  Splitting the Results 45
  Producing Aggregation Statistics 47
  Listing Unique Values 48
  Using Time-Based Functions 50
  Event Order Functions 52
 Eventstats and Streamstats 52
  Eventstats 52
  Streamstats 54
 Using Top and Rare 56
 Chart 58
 Eval 62
  Eval Expressions 63
  Calculating 63
  Converting 64
  Formatting 66
  Rounding 67
  Performing Conditional Operations 67

Creating Visualizations	69
Switching the Type of Visualization	72
Line Chart	72
Area Chart	73
Column Chart	74
Bar Chart	74
Pie Chart	75
Plotting Multiple Data Series	76
Key Takeaways	78

Chapter 3: Using Time-Related Operations 81

Splunk and Time	81
A Note About Time Zone	83
Timechart	84
Specifying Time Span	85
Using Aggregation Functions	87
Using Split-by Fields	89
Basic Examples	91
Additional Useful Tips	94
Retrieving Events in Time Proximity	95
Using the date_time Fields	97
Using Time Modifiers	99
Advanced Examples	102
Comparing Different Time Periods	102
Comparing the Current Day with Average of the Past 30 Days	106
Using Time Arithmetic	109
Key Takeaways	110

Chapter 4: Grouping and Correlating 113

Transactions	114
Using Field Values to Define Transactions	114
Using Strings to Define Transactions	117

TABLE OF CONTENTS

    Using Additional Constraints 120

    What Happens to the Fields in a Transaction? 122

    Finding Incomplete Transactions 124

  Subsearches 126

    Constructing a Subsearch 126

    Problems with Subsearches 128

  Join 129

    Constructing a Join 130

    Problems with Join 132

  Append, Appendcols, and Appendpipe 132

    Append 132

    Appendcols 135

    Appendpipe 136

  Key Takeaways 138

Chapter 5: Working with Fields 141

  Why Learn About Fields? 141

    Tailored Searches 141

    Insightful Charts 144

    Flexible Schema 146

    Index-Time vs. Search-Time Fields 148

  Automatically Extracted Fields 149

    Default Fields 149

    Internal Fields 150

    Fields Extracted Through Field Discovery 152

  Manually Extracting Fields 153

    Using Field Extractor Wizard 154

    Using Field Extractions Menu 160

    A Primer on Regular Expressions 165

    Using Rex Command 168

Using Fields	171
Filtering	171
Sorting	173
Deduping	174
Key Takeaways	175

Chapter 6: Using Lookups 177

Types of Lookups	178
File-Based Lookups	179
Creating a Lookup Table	179
Uploading the Lookup Table File	180
Verifying the Lookup Table Contents	182
Using Lookups	183
The Lookup Command	183
Maintaining the Lookup	186
Using the outputlookup Command	187
Lookups Best Practices	188
Creating Automatic Lookups	189
Key Takeaways	190

Chapter 7: Advanced SPL Commands 193

predict	193
kmeans	195
cluster	196
Outlier	198
fillnull and filldown	200
convert	202
Handling Multivalued Fields	204
makemv	204
nomv	207
mvexpand	208
mvcombine	209

TABLE OF CONTENTS

- mvcount 210
- mvindex 211
- mvfilter 211
- mvfind 212
- mvjoin 213
- mvsort 213
- split 214
- Extracting Fields from Structured Data 214
  - spath 216
- Key Takeaways 220

Chapter 8: Less-Common Yet Impactful SPL Commands 223

- iplocation 223
- geostats 225
- untable 226
- xyseries 227
- bin 228
- tstats 230
  - Using where and by Clause 230
  - Querying Against Accelerated Data Models 231
  - Splitting by _time 233
  - Caveats with tstats 234
- eval coalesce Function 234
- erex 235
- addtotals and addcoltotals 237
- loadjob 238
- replace 239
- savedsearch 240
- Key Takeaways 241

TABLE OF CONTENTS

Chapter 9: Optimizing SPL 243

Factors Affecting Performance of SPL 243
Quantity of Data Moved 244
Time Range of the Search 244
Splunk Server Resources 245

Optimizing Searches 245
Use Fast or Smart Search Mode 245
Narrow Down Time Range 246
Filter Data Before the First Pipe 247
Use Distributable Streaming Commands Ahead in the Pipeline 247

Best Practices for Scheduling Searches 249
Stagger Your Searches 249
Use cron for Maximum Flexibility 250
Utilize Schedule Window Setting 252

Useful Splunk Knowledge Objects to Speed Up Searches 252
Accelerated Reports 253
Summary Indexes 254
Accelerated Data Models 255

Using Job Inspector 257

Key Takeaways 260

Index 263

About the Author

Karun Subramanian is an IT operations expert and a Splunk certified architect. He is committed to helping IT organizations implement world-class observability by making use of machine-generated data. His IT career has spanned more than two decades, ranging from systems administrator to software engineer to IT director. Possessing deep expertise of the Splunk platform, he has assisted teams to solve complex problems in the area of DevOps, security, and business analytics. He has worked in engineering roles for firms including Wells Fargo Bank, Express Scripts, Federal Reserve Bank, and Optum. He has authored several courses on Pluralsight and has delivered trainings on the O'Reilly Live Training platform.

About the Technical Reviewer

Paul Stout is a principal architect for Kaiser Permanente managing Splunk content and development for Kaiser's Technology Risk Office. Paul has written and published several Splunkbase apps including the initial versions of Splunk for ServiceNow, Splunk for Salesforce.com, Splunk for Okta, and the popular WebGL globe visualization. He has worked for or with Splunk for 9 years including a 4-year tour at Splunk managing the company's deployment of Splunk at Splunk, taking value created and lessons learned back to the engineering and sales teams, and sharing Splunk value at conferences, university lectures, and as a deputized sales engineer. He has used Splunk in production use cases across IT, security, engineering, marketing, sales, legal, and HR. In addition to professional Splunk development, Paul has Splunked Corvettes, Cadillacs, power grids, and even his own body and environment through Fitbits, nutrition tracking, and home automation/sensor machine data. He began his technical career developing ETL tools during an active duty tour in the US Air Force unifying reporting from 17 distinct tracking systems, worked as a web developer for a database company, and has held several Splunk consulting/contracting positions. Paul holds a master of business administration and a second-degree black belt in wushu.

Acknowledgments

I would like to thank my dear friend (and running buddy) Steve Buchanan for introducing me to Apress, without whom this book wouldn't have existed. I would like to thank Joan Murray, the acquisition editor of this book, for providing unparalleled advice in writing. I'm greatly indebted to Jill Balzano for coordinating the entire book-writing process, and for keeping me on time. Many thanks to Laura Berendson, the development editor, for ensuring the content is of top quality. Finally, my sincere thanks to Paul Stout for reviewing my work and providing his valuable feedback.

Introduction

Splunk has emerged to be the prominent *data-to-everything* platform. It is evident from the penetration it has made in organizations; you can find Splunk being used in IT operations, business analytics, DevOps, AIOps, and security. Whether you are a software developer, a NOC analyst, or a cybersecurity specialist, Splunk can help find answers from massive amounts of data, fast. The Splunk platform achieves this by collecting and indexing – the process by which raw data is converted into searchable events – virtually from any data source, at scale.

Based on my experience, the biggest hurdle a Splunk user encounters when starting to use the platform is the steep learning curve of Splunk's Search Processing Language (SPL). SPL is the core of Splunk platform. It is the language you use to query the platform to find answers. With its unique Unix *pipe* and SQL-like syntax, armed with more than 140 commands, SPL provides a robust language for searching, analyzing, and creating reports from your data. Shockingly, while there are scores of Splunk books available, none of them is fully dedicated to SPL. And this book aims to fill that gap.

Who This Book Is For?

A Splunk user is a person who is trying to solve problems using the machine-generated data. This person can be a software architect, application developer, data scientist, QA analyst, project manager, support engineer, forensic engineer, vice president of marketing, or even CEO of a corporation. The sole focus of the book is learning to use SPL. While users with all experience levels will benefit from this book, it is especially useful for Splunk beginners and intermediate users.

INTRODUCTION

What This Book Is Not?

This book does not cover the following topics: Splunk knowledge objects such as dashboards and alerts, Splunk systems administration, Splunk data administration, and architecting Splunk. These topics deserve their own books, and there are many books in the market that cover these.

Chapters at a Glance

The book is structured in a way that is practical for users to follow. Rather than alphabetically listing all SPL commands and explaining them, it focuses on the functionality of SPL. For example, after the introduction, the book dives right into calculating statistics, the most common type of work done using SPL.

Chapter 1, "Introducing the Splunk Platform," introduces the Splunk platform and the problems it solves. It discusses the architecture of the platform at a high level. It then introduces the Search Processing Language (SPL) along with its syntax and usage. This chapter also walks you through the Splunk search interface. At the end of this chapter, you will have written your first SPL query yourself.

Chapter 2, "Calculating Statistics," dives right into calculating statistics, an important function of SPL. It explains the all-powerful `stats` command with plenty of examples. It also covers the `chart` command. This chapter then dives into another useful command `eval` and discusses its most useful functions. At the end of this chapter, you will have thorough knowledge of using `stats`, `chart`, and `eval`.

Chapter 3, "Using Time-Related Operations," reveals how you can generate insightful results from your machine-generated data using time. It discusses the command `timechart` in detail and provides various examples. It also provides a few advanced examples such as comparing two different time frames and using the `timewrap` command.

Chapter 4, "Grouping and Correlating," discusses a very practical use of SPL, grouping conceptually related events to make sense. It explains the powerful `transaction` command with many examples. It shows the various constraints you can use to group events. It also introduces `subsearches` and moves on to cover `join` and `append` commands.

Chapter 5, "Working with Fields," shows how you can extract fields from your raw data. It shows how to identify automatically extracted fields and walks through the field extractor wizard for manually extracting fields. It then introduces regular expressions and dives into the rex command. Finally, it discusses the practical uses of fields such as deduping, filtering, and sorting.

Chapter 6, "Using Lookups," takes a little detour and discusses lookups. It walks through using a CSV file as a lookup table to enhance the search results. It explains the use of lookup and inputlookup commands. It also shows how to create and maintain a lookup table using the outputlookup command.

Chapter 7, "Advanced SPL Commands," introduces some of the advanced commands such as predict, kmeans, and cluster. It provides examples for commands such as convert and outlier. It also explains the various ways to handle multivalued fields by using commands such as mvcombine and mvexpand. Finally, it teaches various eval functions such as mvcount and mvfind that help with multivalued fields.

Chapter 8, "Less-Common Yet Impactful SPL Commands," discusses commands such as geostats, iplocation, and tstats. It explains how machine-generated data can be plotted in a map. In addition, it shows how to use erex to automatically generate regular expressions.

Chapter 9, "Optimizing SPL," deals with improving performance of your SPL queries. It explains the factors affecting search performance and shows how to use job inspector, an important tool to understand execution costs of various components of your search. It also discusses the best practices for scheduling searches.

To get the most out of this book, you must actually try the SPL examples yourselves. In addition, refer to Splunk's own documentation at docs.splunk.com for deepening your knowledge. Constant practicing, either in real-world or in your lab environment, will make you increasingly comfortable with writing (and finding answers) using SPL. Let's begin.

CHAPTER 1

Introducing the Splunk Platform

Splunk is the world's leading operational data intelligence platform. It is used by software developers, site reliability engineers, DevOps engineers, data scientists, security professionals, and business analysts. It can parse and index large quantities of unstructured machine data (in many cases, hundreds of terabytes per day). Data is ingested into Splunk using varieties of means, and I find one big advantage being that there is no data source that cannot be ingested into Splunk. The Splunk platform can help to reduce incidents, improve observability, shorten MTTR (mean time to repair), and reveal deeper insights. With a little bit of learning, you will also find Splunk to be a joy to work with. I've personally interacted with hundreds of Splunk users who love the versatility and richness of the platform. The Splunk platform provides an end-to-end solution to manage your machine data, regardless of where it lives or how big it is.

In this chapter, I'm going to introduce you to the Splunk platform and get into the problems that it solves. I'll show you how Splunk can help organizations of all sizes make effective use of machine data. I'll provide an overview of Splunk platform architecture and walk you through various use cases of Splunk. Then, I'll move on to introduce Search Processing Language (SPL), the language of the Splunk platform. I'll show you how to navigate the Splunk Web interface, which is the primary interface you will be using to run searches. Finally, I'll have you run your first SPL query.

Machine Data

Splunk enables you to collect, organize, and make use of machine data, at scale. But first, let's define what machine data is.

CHAPTER 1 INTRODUCING THE SPLUNK PLATFORM

What Is Machine Data?

Simply stated, machine data is the digital exhaust of any computing system. The most common example of machine data is the log files generated by software applications, but it is not limited to log files. Machine data includes monitoring metrics, traces, and events. Figure 1-1 shows the four main categories of machine data. We'll take a look at each of these categories in more detail.

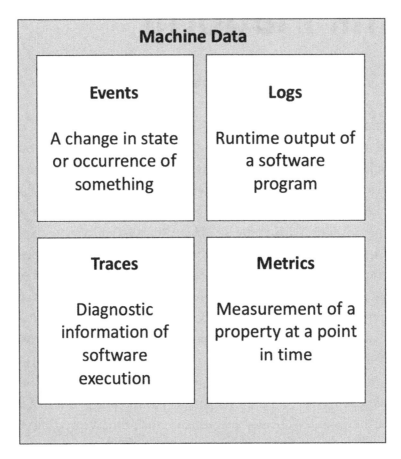

Figure 1-1. *The four categories of machine data*

Events

Events indicate an activity that results in some state change in a system. They are extremely important to understand major changes as the system operates. The way events are stored and retrieved greatly varies by the system. A classic example of events can be easily seen by using Windows Event Viewer, a tool that lets a user view events logged by the Windows operating system.

Listing 1-1. A sample Windows event

```
4954: Windows Firewall Group Policy settings have changed. The new settings
have been applied.
```

Note In Splunk, an event is a single piece of indexed data. Data ingested by Splunk is broken into events and stored in the index.

In general, event data can be structured or semistructured though it depends on the system that generates the events.

Logs

Logs are data generated by software as a byproduct during operation. Often, logs are used to convey pertinent information about activities happening during runtime. Logs can be very useful to troubleshoot operational issues. Logs can also be used to store audit information in order to be used for security compliance. Logs are generally stored as text files. Most industry standard products such as Apache HTTP Servers generate well-known log file formats such as *access.log* and *error.log*. However, custom software applications typically have log files that are determined by the application development team. For instance, the log file of an application that runs batch jobs may be called *jobActivity.log*.

CHAPTER 1   INTRODUCING THE SPLUNK PLATFORM

Listing 1-2. A sample web access log

```
172.17.0.1 - admin [31/Dec/2019:12:28:15.005 +0000] "GET /en-US/
splunkd/__raw/servicesNS/admin/splunk_monitoring_console/alerts/alert_actions?
output_mode=json&search=(is_custom%3D1+OR+name%3D%22email%22+OR+name%
3D%22script%22+OR+name%3D%22lookup%22)+AND+disabled!%3D1&count=
1000&_=1577795294532 HTTP/1.1" 200 2204 "-" "Mozilla/5.0 (Macintosh;
Intel Mac OS X 10_14_6) AppleWebKit/537.36 (KHTML, like Gecko)
Chrome/79.0.3945.79 Safari/537.36" - 67c449b576757e486a63f4b13aa0f313 4ms
```

Logs make up the biggest portion of the machine data ecosystem in organizations. They are typically unstructured, and because of that, they are difficult to index.

Traces

Traces are diagnostic information produced by software as it executes. They are extremely valuable for analyzing the end-to-end flow of a transaction. They typically contain information such as contextual data, runtime parameters, and program stack traces. Due to the rich set of information they contain, traces are often used to diagnose complex problems that traverse multiple systems. Traces are generally stored as text files.

In software development, a trace can also mean a stack trace, which shows the execution path of a block of code. They are useful to pinpoint lines of code that could be causing an error. A sample java stack trace is shown in the following.

Listing 1-3. A sample java stack trace

```
java.net.ConnectException: Connection refused
        at java.net.PlainSocketImpl.socketConnect(Native Method)
        at java.net.PlainSocketImpl.doConnect(PlainSocketImpl.java:351)
        at java.net.PlainSocketImpl.connectToAddress(PlainSocketImpl.
        java:213)
        at java.net.PlainSocketImpl.connect(PlainSocketImpl.java:200)
        at java.net.SocksSocketImpl.connect(SocksSocketImpl.java:432)
        at java.net.Socket.connect(Socket.java:529)
        at java.net.Socket.connect(Socket.java:478)
        at java.net.Socket.<init>(Socket.java:375)
        at java.net.Socket.<init>(Socket.java:189)
        at TCPClient.main(TCPClient.java:13)
```

The content and format of the trace files are dependent on the software that creates them. Because the trace files can get very large in a short period of time, they may be produced on an ad hoc basis.

Metrics

Metrics are numeric measurements of a property, typically collected at regular time intervals. They can be very useful for monitoring and trend analysis. Due to the numerical nature of metrics, they are the easiest and most efficient to ingest. I would go the extent to say that metrics are the gold standard of monitoring data.

Listing 1-4. A sample cpu metric

```
cpu_percentageBusy,cpu=cpu-total,dc=sf-13,host=web23a
value=9.2073457085694689
```

Time-Series Nature of Machine Data

Each event in the machine data is often accompanied by a timestamp. Storing time-series data significantly differs from traditional database storage. With time-series information, data is typically stored in time-based bins (or buckets). Further, data management is strongly time oriented. For instance, data retention policies can be implemented to meet various user needs based on the age of the data.

The Value of Machine Data

It is easy to overlook the value you can derive from machine data. However, I've seen scores of teams solving varieties of problems using machine data. For example, machine data can be used to implement real-time error rate monitoring of a service. Here are the four major areas where you can utilize machine data.

IT Operations and Monitoring

This is perhaps the most popular use case of machine data. By collecting and organizing all of your organization's machine data in one central place, you will greatly improve the MTTR (mean time to repair). For example, by continuously monitoring the error rate of your application, you can alert the on-call about a potential outage long before a major outage occurs. Figure 1-2 shows a graph that plots Raw Error Rate against time.

CHAPTER 1 INTRODUCING THE SPLUNK PLATFORM

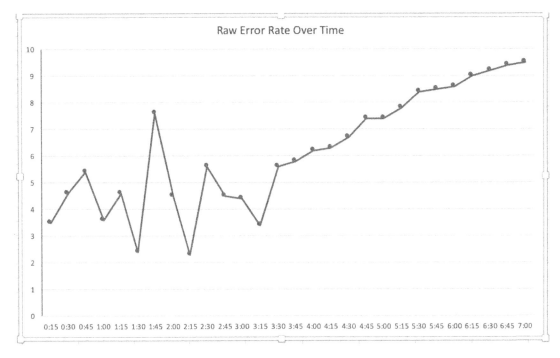

Figure 1-2. *Raw Error Rate of an application over time, which is derived from machine data*

Security and SIEM

Machine data is a treasure trove for uncovering fraud and illicit security attacks against your systems. Many organizations use machine data in the form of audit logs for identifying malicious login attempts. The web access log can help you identify a DDoS (Distributed Denial of Service) attack by looking for patterns in the *client IP* field. With a powerful query language such as Splunk Search Processing Language (SPL), an analyst can run effective ad hoc queries and correlate events across many sources. Further, by collecting and indexing audit logs, you can create useful reports to meet compliance requirements.

Business Analytics

Using machine data in business analytics is relatively a new user case compared to IT analytics and security. Traditionally, business analytics is performed using transactional data stored in huge data warehouses such as Teradata. But machine data contains several bits of useful information that are often overlooked though very handy for business analytics. For instance, you can measure the geographical distribution of your sales volume using web server and application server log files. And with the adequate machine data, you can also automatically discover business process flows.

AIOps

AIOps (Artificial Intelligence for IT Operations) is fast emerging as a disruptive technology. AIOps comprises related technologies such as machine learning, predictive analytics, and anomaly detection. Collecting and indexing all of the machine data in your organization opens doors for many innovative use cases using AIOps.

Note Splunk provides Machine Learning Toolkit, an app that you can install on top of Splunk Enterprise to use machine learning algorithms on your data.

The Shortcomings of Machine Data

If machine data is so useful, then why aren't all organizations making the full use of it? There are four reasons why machine data does not get the attention it deserves.

Size

Machine data can be extremely large in terms of size. In organizations of any size, machine data can easily grow to several terabytes per day (especially when collecting logs from network devices). In web-scale companies, it can even reach petabytes. Collecting and ingesting this much data requires well-designed software platforms such as Splunk Enterprise. Traditional data warehouses don't help much as they are not designed for indexing time-series data. Big data platform such as Hadoop requires complex map-reduce jobs to retrieve data as they are primarily used for batch loads.

Speed

Machine data has enormous velocity. It is not uncommon to see millions of events per second, especially when you collect metrics from operating systems and network devices. This makes it hard for reliability collecting and indexing them. Once again, a well-designed platform such as Splunk will need to be deployed if you want to make full use of machine data.

Structure

Logs, the largest machine data category, largely tend to be unstructured. This makes it difficult to parse and index. Traditional data stores require the data to be either structured or, at least, semistructured. Splunk platform solves this problem by employing *schema-on-read* technology. During parsing and indexing, Splunk extracts just a few metadata fields such as timestamp, source, sourcetype, index, index time, and host. All other field extractions happen during the search-time. This means that you only need to have some high-level idea about the machine data being ingested (the most important ones being line-break pattern and timestamp). In *schema-on-write* however, you need to fully define the fields in the machine data ahead of indexing.

Distribution

Machine data is everywhere. Server log files, application traces, IoT devices logs, sensor outputs, and voice call records are examples of machine data. Since the data is widely spread out and often difficult to collect, many organizations don't even consider the possibility of collecting and ingesting all of machine data in one central system. Recent explosion in containers and cloud platforms make it even more complex.

The Splunk Operational Data Intelligence Platform

Splunk is the industry leader in operational data intelligence. With its powerful Search Processing Language (SPL), which is the focus of this book, and rich set of visualizations, Splunk aids in solving the problems with the machine data mentioned earlier. In addition, the Splunk platform employs an extensible architecture. This has paved ways for thousands of community-developed Splunk apps, which are a set of configuration files, program files, and visualizations that extend Splunk's capabilities.

Primary Functions of Splunk

There are five functions of Splunk. Each of these functions addresses a specific area of handling machine data. Let us review them in detail.

Figure 1-3 shows the five functions of Splunk.

CHAPTER 1 INTRODUCING THE SPLUNK PLATFORM

Figure 1-3. *Five main functions of Splunk operational data intelligence platform*

In the *Collect and Index* function, Splunk offers several ways to receive data from machine data sources. The most common way is using *Splunk Universal Forwarder*, which is a piece of software installed on the machine data hosts. There are many other ingestion mechanisms such as *Splunk HTTP Event Collector* and *Splunk DBConnect*. The collected data is parsed and indexed in indexes, which are flat files that consist of raw data and time-series index files.

In the Search and Investigate function, Splunk provides Search Processing Language (SPL) to query the indexed data. With its powerful SQL-like and Unix pipe syntax, SPL provides unprecedented flexibility when it comes to slicing and dicing your machine data. SPL provides more than 140. In addition, because of schema-on-read, you are able to manipulate, group, and correlate data on the fly. SPL provides a family of commands known as transforming commands using which you can generate reports and visualizations just with one or two commands.

In the *Add Knowledge* function, Splunk offers a variety of tools such as lookups, field extractions, tags, event types, workflow actions, and data models. These tools help you to make sense of your machine in the quickest and most effective way.

In the *Report and Visualize* function, Splunk offers the capability to produce stunning reports and dashboards and even schedule them for automatic delivery. Splunk Search Processing Language (SPL) provides a set of commands called *transforming commands* such as `stats` and `timechart` that instantly creates varieties of tables and visualizations based on the search results.

Finally, in the *Monitor and Alert* function, Splunk enables you to monitor your systems and trigger alerts based on the thresholds you set. There are varieties of actions you can perform when an alert is triggered, the most common one being email alerts. Splunk can also call a REST endpoint using HTTP post request. In addition, you can write your own action using Splunk's custom alert action framework.

Architecture of Splunk Platform

Splunk platform is architected in a distributed computing model which enables reliable scaling and performance. At its core, Splunk can be thought of as a search engine for operational machine data. In that regard, it is not very different from traditional search engines like Google. In fact, the inverted index that Splunk builds as it ingests operational data is not very different from a search engine index. While searching, Splunk employs a map-reduce algorithm very similar to search engines and big data platforms like Hadoop. Figure 1-4 depicts a high-level architecture of Splunk.

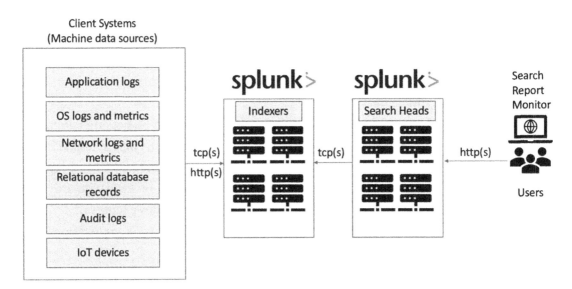

Figure 1-4. *High-level architecture of Splunk operational data intelligence platform*

The following are three major components of Splunk platform:

1. Indexer
2. Search head
3. Forwarder

Let us take a look at each of the preceding components in detail.

Indexer

The indexer is the core Splunk Enterprise process which converts raw data into searchable events and stores them in indexes. During indexing, the indexer breaks the raw data into events, extracts timestamp (and few other metafields), and writes the data to the disk. It is also the process that searches and retrieves the indexed data. In Unix systems the indexer process is named *splunkd*. In a distributed environment, the indexer is also called *search peer*. Many indexers are grouped in a cluster, and data is replicated among cluster members.

Note Once data is indexed, it cannot be modified.

Search Head

The search head handles the searches run by the users. It is a Splunk Enterprise process that employs map-reduce architecture. It distributes search requests (the map phase of map-reduce) to a group of indexers (also known as search peers) where the searches get executed. Search head receives the search results and merges them (the reduce phase of map-reduce) before sending the results back to the user. Search heads can also be grouped in a search head cluster. Search head clusters provide high availability and load balancing. They also help in controlling access to the indexed data. Typically, when you log in to the Splunk Web interface, you are logging on to the search head.

Note It is possible to have the same splunkd process assume both indexer and search head roles. This setup is common in dev/lab environments.

Forwarder

The forwarder is a Splunk Enterprise software process that collects data from the machine data host. The most common type of forwarder is the Splunk Universal Forwarder which is typically installed in Unix and Windows operating systems. Based on the configuration, universal forwarder collects and sends machine data to indexers. Universal forwarders can also send the machine data collected to a special type of forwarders called *heavy forwarders*. Unlike universal forwarders, heavy forwarders parse the data before sending them to indexers. Heavy forwards can also optionally index the data locally.

There are other types of Splunk Enterprise components such as deployment server, search head deployer, cluster master, and license master. They are typically present in large-scale distributed environments. There are other books that cover the full range of Splunk features. However, as the focus of this book is on the Search Processing Language, these components will not be covered in detail. So let's dive into the language next.

Introducing Splunk Search Processing Language (SPL)

Splunk Search Processing Language (SPL) is the language of the Splunk platform. It is the language used to perform searches against the indexed data. With its versatile SQL-like and Unix pipe syntax, SPL offers a very powerful way to explore and investigate your machine data. You usually run SPL queries using the *Search & Reporting* app that comes by default with Splunk Enterprise. You can also use Splunk REST-API and command-line interface (CLI) to run SPL queries. The search is executed by Splunk Enterprise, and the results are shown in the user interface. By employing certain commands in SPL, you can also generate reports and visualizations of your search results.

Syntax of SPL

The Search Processing Language is a hybrid of SQL (Structured Query Language) and Unix pipe. Just like any other query language, SPL follows certain syntax. When a query has a syntax error, it does not execute, and an error is shown indicating the problem with the syntax. Let us take a look at the general syntax of SPL. An SPL query can consist of any or all of the following.

Commands

Commands are keywords that execute certain functionality in Splunk. A basic (yet very powerful) command in Splunk is `search`. Commands can accept one or more parameters. Here are some examples of commands:

- stats
- eval
- timechart
- table
- fields

Note At the start of an SPL query, the `search` command is implied so you don't need to explicitly type it. The search command must be specified if it comes later in the search pipeline.

Literal Strings

You can search for any string or phrase. If the phrase has spaces, you need to enclose the phrase in double quotes. Here are some examples of searching for literal strings:

- Error
- i/o Exception
- "java.lang.NullPointerException"
- "Order processed successfully"
- "Program quit unexpectedly"

Key-Value Pairs

When Splunk indexes the data, it converts raw data into searchable events. During the process, it extracts certain key-value pairs and stores them as fields. These fields are generally the default fields such as host, source, and sourcetypes. Fields can also be extracted during search-time. You can use these fields to filter your data. Using key-value pairs is one of the most basic yet significant ways of searching your data. The key must be a field in the events. Here are some examples of key-value pairs:

- index = main
- sourcetype = userPortal:app:errorLog
- status = 500
- result = "Order processed successfully"
- log_level = ERROR

Note If the field name contains spaces, use single quotes around them, for example, 'user id' = jsmith.

Wildcard

You can use the wildcard asterisk (*) to match zero or more of any character. This is the only wildcard supported when used in literal strings and as part of key-value pairs. Here are some examples of using wildcard:

- status = 5* ← Retrieves events that have status starting with 5. For example, 500, 503, and 504
- "java.net.Connection*" ← Retrieves events that have the string java.net.Connection followed by zero or more of any character. For example, java.net.ConnectionRefused, java.net.ConnectionTimedOut
- "Login for user*failed" ← Retrieves events that have the string "Login for user" followed by zero or more of any character, followed by the string "failed"

- "Account number:*activated" ← Retrieves events that have the string "Account number:* followed by zero or more of any character, followed by the string "activated"

Note Using the * wildcard as the first character in a string is very inefficient. For example, avoid searching for strings like "*error".

You can also use the `like` function of eval command to perform SQLite-style wildcard matching. The like function supports percent (%) to match multiple characters and underscore (_) to match a single character. We discuss `eval` and `like` in Chapter 2, "*Calculating Statistics*."

Comparison Operators

One of the most powerful ways to filter your data is to use comparison operators to check for conditions. Using comparison operators, you can compare the values of fields. The following operators are supported:

- =
- ==
- !=
- >
- >=
- <
- <=

Here are some examples of using logical operators:

- percentage_utilzation < 50
- response_time >= 500

Note The operators = and == are treated the same when comparing.

Boolean

You can use Boolean operators to add logic to your queries. The following Boolean operators are supported:

- AND (implied if not explicitly specified)
- OR
- NOT

Here are some examples of using Boolean:

- log_level = WARNING OR log_level = ERROR
- result = Failed AND user = jsmith
- output = "Login succeeded" NOT admin
- result = Error OR (result = Pass AND response_time > 500)

Note The Boolean operators must be in uppercase. You can also use parenthesis to force precedence.

Functions

Many SPL commands have functions that you can utilize when using those commands. Functions typically operate on parameters you specify. Here are some examples of SPL functions:

- ...| stats *avg*(response_time). ← Calculates the average response time
- ...| timechart perc95(processingTime) ← Calculates the 95th percentile of processingTime
- .... | eval percentSuccess = round(success_rate,2) ← Rounds the success_rate to two decimal places and stores the result in a new filed percentSuccess

Note eval command calculates an expression and stores the result into a field.

Arithmetic Operators

Arithmetic operators help with performing numerical calculations. They are generally part of an eval expression. Table 1-1 shows the various arithmetic operators you can use.

Table 1-1. *SPL arithmetic operators*

Arithmetic operator	Meaning	Example
+	Sum	… | eval counter = counter + 1
-	Subtract	… | eval price = total – discount
*	Multiply	… | eval tax = income * .20
/	Divide	… | eval dollars = pennies / 100
%	Modulo	… | eval rem = num % 2

The Search Pipeline

SPL utilizes Unix pipe notion. By using the pipe character (|), results of the left side of the pipe can be passed to the right side of the pipe. This provides an excellent way to add logic to your queries. Consider the following SPL statement:

```
index=main sourcetype=secure "Failed password" | top UserName |
fields - percent
```

Figure 1-5 depicts the search pipeline.

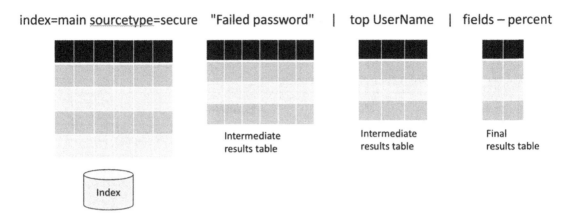

Figure 1-5. *The search pipeline*

As indicated in the preceding figure, as you traverse through the pipeline from left to right, the results are increasingly narrowed (although in some cases, you may manually add new fields by using `eval, lookup,` and subsearches). While troubleshooting why some SPL queries don't return the results you want, it is a good idea to keep removing the segments from right to left to review the intermediate results table.

Note Do not worry if you don't fully understand the SPL query at this point. We will cover the commands such as `top` and `fields` later in this book.

Navigating the Splunk User Interface

You interact with Splunk using *Splunk Web*, a web-based interface that comes with the Splunk product. *Search & Reporting* app (also known as *Search app*) is a preinstalled app that you use to run searches and create reports. It is important to understand all the features of the *Search app* to aid in effective use of Splunk.

Note An app in Splunk is a set of configuration files and/or visualizations installed on Splunk Enterprise platform. App is the primary mechanism by which the basic functionality of Splunk is extended.

Installing Splunk

The best way to learn Splunk is to use your own Splunk environment (preferably nonproduction) as you will be searching the data you are already familiar with. But if you don't have access to a Splunk environment, it is easy to download and install a trial version on your PC, Linux, or Mac. The trial includes all features of Splunk with 500MB/day ingestion limit for 60 days. After 60 days, the license automatically changes to free version at which point certain features like alerting and access control are disabled. You can also use Splunk Cloud Trial which is a SaaS (Software as a Service) offering from Splunk. Use the following links to get started:

Splunk Enterprise download: `https://www.splunk.com/en_us/download/splunk-enterprise.html`

Splunk Cloud Trial: `https://www.splunk.com/en_us/campaigns/splunk-cloud-trial.html`

Simply follow the instructions from the preceding links to install and start up Splunk. The default port for Splunk Web is 8000.

During the install, you would be asked to set up an administrator user and password. By default, the administrator user is called admin. Make sure to keep the credentials for the administrator user safe.

Logging onto Splunk Web

Log on to Splunk Web interface using a web browser. If you had installed the trial version on your PC, Linux, or Mac, the URL to access *Splunk Web* will most likely be the following:

`http://localhost:8000`

At the login screen, provide the administrator user ID and password. By default, the app *launcher* is invoked. *Launcher app* is also preinstalled with Splunk. Figure 1-6 shows the typical home page of the *launcher* app.

CHAPTER 1 INTRODUCING THE SPLUNK PLATFORM

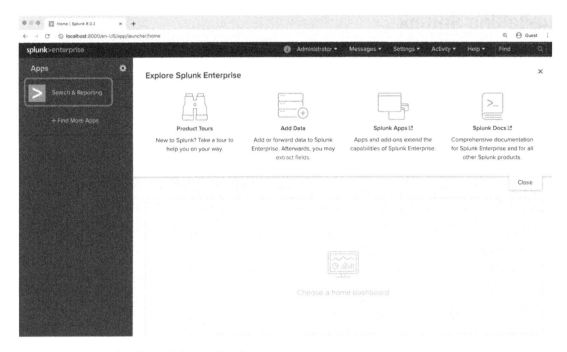

Figure 1-6. *Splunk Web launcher home page*

Note If you are using your organization's Splunk environment, depends on how your Splunk administrator has configured access control, your home page may look different.

As highlighted, click *Search & Reporting* to launch the search interface.

CHAPTER 1   INTRODUCING THE SPLUNK PLATFORM

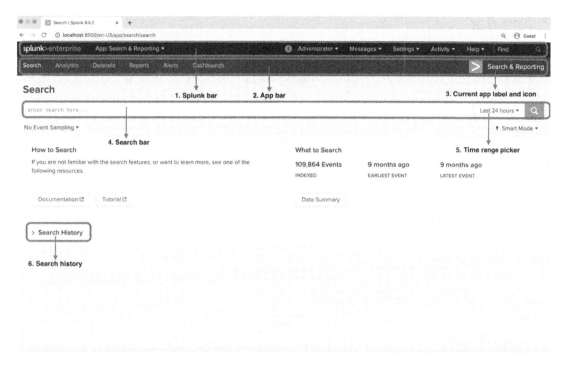

Figure 1-7. Splunk Web Search & Reporting home page

Let us look at the various sections of the *Search & Reporting* home page and their use.

1. Splunk Bar

The black bar at the top is called the *Splunk bar*. This bar has many useful menu options. For example, you can set your default time zone using the Preferences menu located under your username. The Activity menu can be used to monitor the search jobs you have submitted. Figure 1-8 shows the various options available in the Splunk bar.

CHAPTER 1   INTRODUCING THE SPLUNK PLATFORM

2. App Bar

The App bar contains navigation menus available in the current app. This can be customized using configuration files bundled within the app. By default, Splunk comes with *Search & Reporting* app.

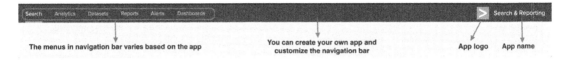

3. App Icon and Label

The current app icon and label show the app that you are currently on. It is a good idea to periodically make sure that you are in the correct app as you navigate through different screens within *Splunk Web*.

4. Search Bar

The search bar provides a text field for you to enter the SPL query. The keywords are color coded as you type. For example, the commands usually appear in blue, and functions are in pink and so on. To execute a search, you can simply press Enter, or click the magnifying glass icon at the right.

5. Time Range Picker

The time range picker enables you to specify the time range for your query. The default time range can be configured by your Splunk administrator. I always suggest setting the time frame as narrow as possible to get the best search performance. You can pick from any of the preset time frames or use advanced options to specify a custom time frame.

6. Search History

You can retrieve your previously run searches using search history. This can come in handy if you don't want to retype your SPL queries. You can also search within the search history using the filter found at the top of the Search History pane.

> **Note** If you are using Splunk in a search head cluster setup, depends on which search head member you land on when you log in, you may not see your previously run searches. By default, search history is not replicated among search head members. An administrator can override this by configuring `conf_replication_include.history=true` in server.conf.

Write Your First SPL Query

At this point, you have learned the basic functionality of Splunk, a high-level understanding of SPL syntax and the search interface. Let's move ahead by actually writing your first SPL query. In this section I'll walk you through uploading Splunk tutorial data, and we'll start querying.

Using Splunk Tutorial Data

While using your own data in your organization is certainly helpful to learn Splunk, to follow along the examples I am going to show you, you will need to install the Splunk Enterprise trial in your PC/Mac and upload the data sets I provide. This way, you can avoid mixing the tutorial data with your organization's data. See the section "Installing Splunk" in this chapter to find the instructions for installing Splunk trial version.

Download Splunk Tutorial Data Zip File

You will need to download the Splunk tutorial data zip file using the following link:

`https://docs.splunk.com/Documentation/Splunk/latest/SearchTutorial/Systemrequirements#Download_the_tutorial_data_files`

> **Note** You can always search Splunk product documentation at docs.splunk.com to locate the Splunk tutorial download files.

The name of the tutorial zip file is *tutorialdata.zip*.
Download the zip file tutorialdata.zip into your PC/Mac. You will be uploading this file into Splunk in the next step.

CHAPTER 1 INTRODUCING THE SPLUNK PLATFORM

Note Do not unzip the downloaded file. You will need the zip file to be uploaded to Splunk.

Add Splunk Tutorial Data Zip File into Splunk

1. Log in to Splunk Web as administrator.
2. Click *Add Data*. See Figure 1-8.

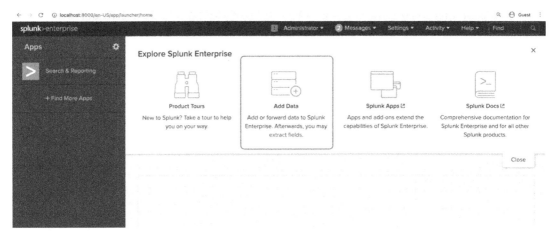

Figure 1-8. *Adding data to Splunk*

Note If you are using your organization's Splunk environment, you may not find the *Add Data* icon. This happens when your Splunk administrator has disabled this option. Use Splunk Enterprise trial if this is the case.

3. Click Upload. See Figure 1-9.

CHAPTER 1 INTRODUCING THE SPLUNK PLATFORM

Or get data in with the following methods

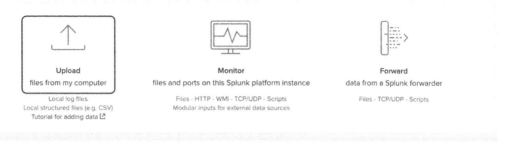

Figure 1-9. Uploading data from your computer

4. Click *Select File*, and choose the tutorialdata.zip file you had earlier downloaded. Make sure the file is in zip format. See Figure 1-10.

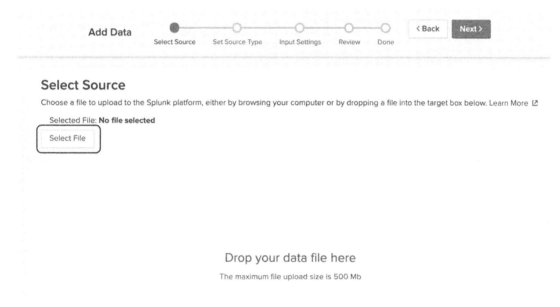

Figure 1-10. Selecting tutorialdata.zip to upload to Splunk

CHAPTER 1 INTRODUCING THE SPLUNK PLATFORM

5. Click *Next*. See Figure 1-11.

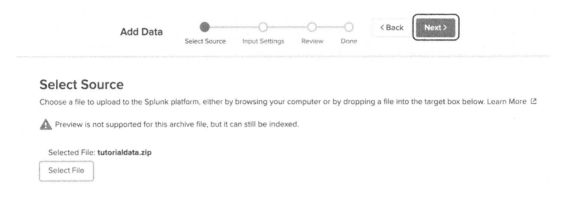

Figure 1-11. *Continuing uploading tutorialdata.zip into Splunk*

6. In the Input Settings screen, choose the *main index* and click *Review* as shown in Figure 1-12.

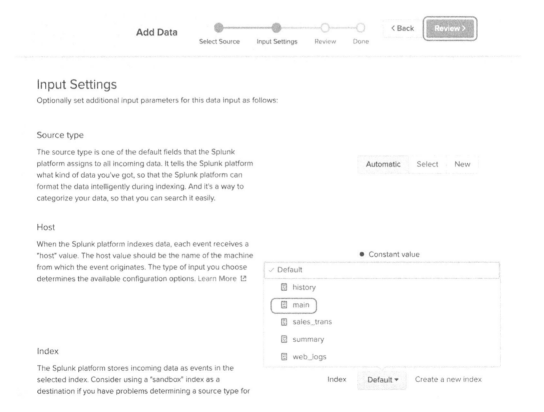

Figure 1-12. *Selecting target index while uploading tutorialdata.zip*

CHAPTER 1 INTRODUCING THE SPLUNK PLATFORM

7. Click Submit in the Review screen. Note that the hostname defaults to the hostname of your computer. See Figure 1-13.

Figure 1-13. *Reviewing the settings before starting the indexing process*

8. Splunk uploads and indexes the data. If there are no errors, you see a success message as shown in Figure 1-14. Click the *Splunk Enterprise* icon at the top left to go back to the home screen.

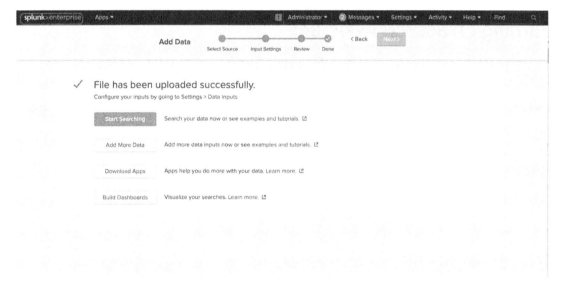

Figure 1-14. *Finishing up the upload process*

27

CHAPTER 1   INTRODUCING THE SPLUNK PLATFORM

9. Launch the *Search & Reporting* app as shown in Figure 1-15. Now you are ready to search.

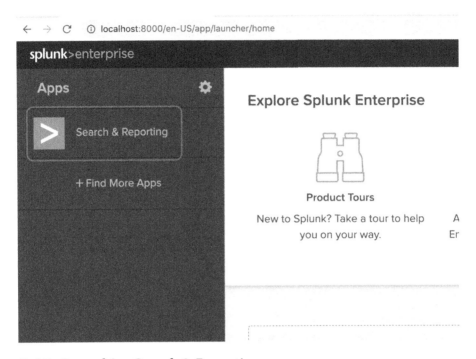

Figure 1-15. *Launching Search & Reporting app*

Turning on Search Assistant

At this point I suspect you are eager to run your first search, but let me first suggest an important setting that will aid you in searching. The *Search Assistant* is a feature of Splunk that listens to every keystroke as you type the SPL and provides suggestions, help, and examples of commands. For example, as you type, the Search Assistant will show the possible SPL command matches. You can easily enable Search Assistant by updating your preferences as described in the following:

1. In the *Splunk bar*, click the down arrow right next to your username and click the *Preferences* menu. See Figure 1-16.

CHAPTER 1   INTRODUCING THE SPLUNK PLATFORM

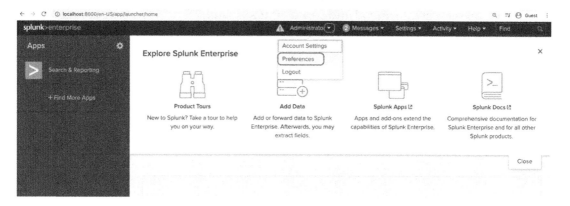

Figure 1-16. *Preferences menu*

2. In the Preferences menu, click the *SPL Editor* tab and select *Full Search Assistant*. In addition, enable *Line numbers* and *Search auto-format* as shown in Figure 1-17. Click *Apply*.

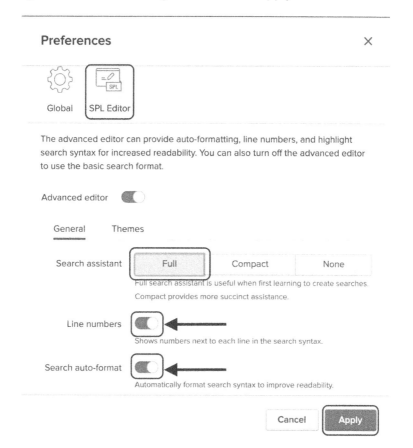

Figure 1-17. *Turning on Search Assistant*

29

You are all set. Let's go ahead and run your first search.

Search Modes

You can run the search in three modes, namely, *fast*, *smart*, and *verbose*. You can choose one of them by using the drop-down menu at the right side of the search bar, under the time picker. The default search mode is *smart*. See Figure 1-18.

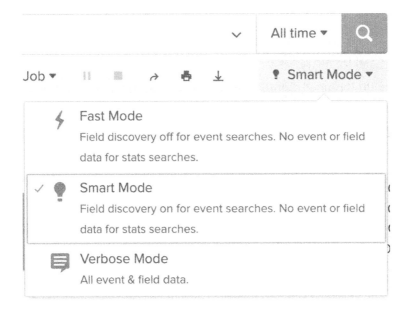

Figure 1-18. Selecting the search mode

Let's go more into depth on the significance of these modes and what they mean in your search.

Fast Mode

Fast mode favors performance. In this mode Splunk does not waste time in discovering fields (field discovery). It only extracts fields that are present in your SPL query. Further, when you use a transforming command such as *stats* to generate a report, fast mode does not generate the event details. Instead it jumps straight to Statistics or Visualization tab.

Verbose Mode

Verbose mode favors thoroughness over performance. It discovers all possible fields. When you run a transforming command such as stats to generate a report, verbose mode generates the events list and timeline. This may be useful in some scenarios. Because of the performance hit (in some cases, as much as ten times), use verbose mode with caution.

Smart Mode

This is the default mode, and I recommend you run your searches in this mode. Smart mode provides a blended approach between fast mode and verbose mode. With smart mode, automatic field discovery is turned on. In this, it is similar to the verbose mode. But when you use a transforming command such as stats to generate a report, smart mode does not generate the full events list. In this, it is similar to the fast mode.

Run the Search

Now that you have turned on the *Search Assistant*, and set the search mode, it's time for you to run the search. From the Search & Reporting home page, type the following SPL into the search bar:

```
index=main
```

Set the time frame to *All time* by clicking the *Time picker* as shown in Figure 1-19.

CHAPTER 1 INTRODUCING THE SPLUNK PLATFORM

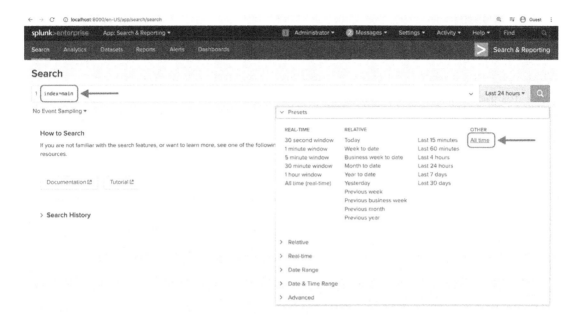

Figure 1-19. Choosing time frame while running the search

To run the search, you either click the *Search* icon at the right corner of the Search bar or simply click anywhere in the search bar where you typed your query, and press *Enter*.

Note   In practice, you should always specify as narrow of a time frame as possible for the best performance. However in this case, we choose *All time* to ensure all of tutorial data is retrieved.

What Happens When You Run the Search?

When you run your search, Splunk goes to work. It validates the SPL you typed and executes your query for the time frame you have selected. In our example, it searches the index named *main* for *All time* and brings back the events to the user interface. The search results are shown in reverse chronological order (newest data first). The search may take a few seconds to run, and the resulting screen will look like Figure 1-20.

CHAPTER 1   INTRODUCING THE SPLUNK PLATFORM

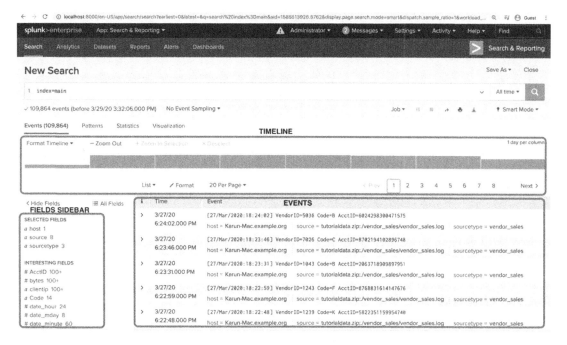

Figure 1-20. Search results

Congratulations! You have just taken your first step toward immersing yourself into the fascinating world of machine data and Splunk.

Before we go deeper, I would like to point out the various sections and menus that just became available after you ran the search. They will come in very handy to make full use of Splunk.

Timeline

The timeline provides a graphical view of the number of events returned from the search. It can be a quick way to identify any spikes in the number of events. For example, if you are searching for a specific string such as *java.lang.NullPointerExeption*, glancing at the timeline can provide a quick way to identify any spikes.

When you click on any area of the timeline, the Events tab automatically zooms in based on the selected time range. You can also drag on the timeline to select a range of time. If you double-click a specific bar in the timeline, Splunk reruns the search with the new time range.

You can hide the timeline completely if you wish to do so. Simply click the *Format Timeline* button and choose *Hidden*. See Figure 1-21.

33

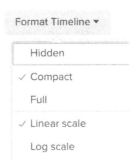

Figure 1-21. *Customizing timeline*

By default, *Compact* timeline is selected. You can choose *Full* timeline which will show x- and y-axis values. In addition, you can choose between *Linear scale* and *Log scale* for the display.

Fields Sidebar

The fields sidebar is where you will find automatically and manually extracted fields. Fields are searchable key-value pairs that Splunk extracts from your data. Fields are extremely useful to tailor your searches. We'll look at fields in depth in a later chapter.

There are two sections of the fields sidebar, *Selected Fields* and *Interesting Fields*.

Selected fields are a set of configurable fields that are displayed for each event. Host, source, and sourcetypes are three default selected fields. Since these are very important fields, let me quickly describe these fields.

Host is the device from which the data originated. Typically, this is the server on which the Splunk Universal Forwarder is installed. In some ingestion mechanisms such as HTTP Event Collection, the hostname can be specified by the client.

Source is the name of the file from which the data has been ingested. Source can also be a stream or other input types.

Sourcetype denotes a specific data type. Splunk automatically recognizes many popular log files (e.g., Apache web server access logs). You can also create custom sourcetypes.

Interesting fields are fields that are present in at least 20% of the events in the result set. These fields could be automatically extracted by Splunk, or manually extracted by you.

You can click on any of the fields to bring up the field menu. From this menu, you can see the top ten values of the selected field, run a quick report, or select the field to be a Selected field. See Figure 1-22.

CHAPTER 1   INTRODUCING THE SPLUNK PLATFORM

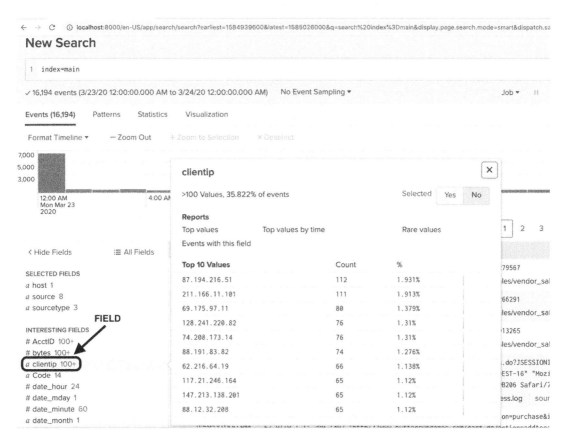

Figure 1-22.   *Fields sidebar*

Event Details

Perhaps the most important section of the results screen is the event details. This is where the raw data is shown as events. Splunk converts raw data into events during the indexing process. An event is a piece of data with a timestamp. Each event also shows the Select fields. See Figure 1-23.

Note   The SPL command *transaction* can group multiple events based on constraints you specify. For example, you can group all events related to a one user session in a web event log.

CHAPTER 1   INTRODUCING THE SPLUNK PLATFORM

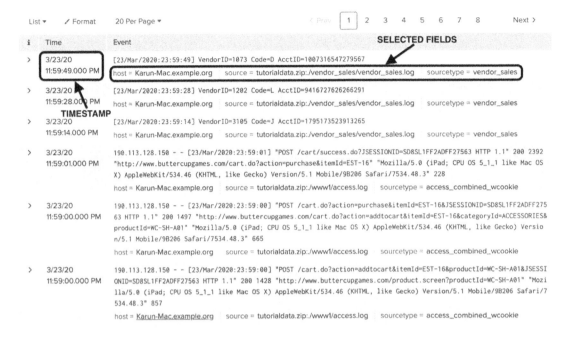

Figure 1-23. *Event details*

As you hover over the events details area, Splunk highlights keywords that you can add to your search by clicking and choosing *Add to search*. You can expand a single event by clicking the arrowhead at the left. See Figure 1-24.

CHAPTER 1   INTRODUCING THE SPLUNK PLATFORM

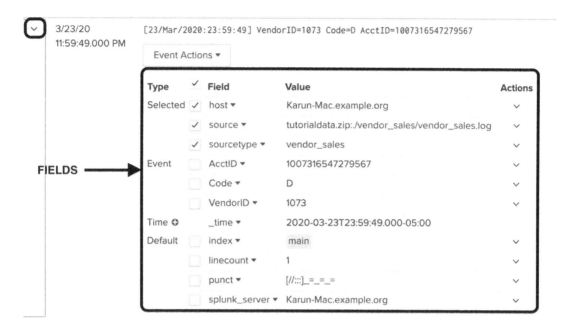

Figure 1-24. *An expanded event*

Key Takeaways

In this chapter, you learned about the basics of the Splunk platform and how it solves many problems with log files. I discussed the high-level architecture of Splunk and explained the most common use cases. I introduced Search Processing Language (SPL), the query language used to retrieve data from Splunk. I also walked you through the Splunk Web interface. Finally, you ran your first SPL query. Few key points to remember from this chapter

1. Splunk can ingest any type of text data, structured, unstructured, or semistructured.

2. Splunk utilizes the concept of schema-on-read in which most of the structuring of the machine data happens at search-time.

3. Splunk organizes and stores data in indexes.

4. Search Processing Language (SPL) is the query language used to retrieve data from Splunk. SPL syntax is a hybrid of SQL and Unix pipe.

5. While there are 140+ commands in SPL, mastering a few key commands is more than enough to make full use of Splunk.

6. While using Splunk Web, turn on full search assistant for a rich experience while searching.

7. Prefer smart mode while searching.

8. The time range you specify for a search will have the biggest influence on search performance.

Now that you have learned the basics of Splunk platform and SPL, in the next chapter, "Calculating Statistics," we'll dive right into useful SPL commands by learning how to calculate statistics from your machine data.

CHAPTER 2

Calculating Statistics

Welcome to the world of machine data, Splunk, and statistics. You are about to see the massive power of Splunk SPL and how it turns data into answers. The Splunk platform can ingest and index massive amounts of machine data and make them easily and readily searchable. Searching your data is an excellent use case of Splunk, but you can do much more. For example, with one simple query, you can use the power of SPL to calculate the average, maximum, minimum, and 95th percentile response times of your application over a period of time. You can funnel large amounts of raw events into Splunk and get meaningful statistics as output, within a few seconds in many cases.

SPL has a rich set of commands to calculate statistics from your machine data. In practice, you will often use these commands to develop reports, alerts, and dashboards. In this chapter, I'm going to explain the most commonly used statistical commands. I will also cover how to create basic visualizations using these commands.

Stats

`Stats` is a SPL command that falls into the category of *transforming commands*. A transforming command takes the results of a search and orders them into a table which can be used for statistical purposes. In order to produce visualizations such as pie charts and bar charts, you must use a transforming command. Popular examples of transforming commands include the following:

```
stats
chart
timechart
top
```

In this section, we are going to focus on `stats` command.

CHAPTER 2 CALCULATING STATISTICS

A Quick Example

Consider the following events:

```
04-21-2020 23:47:19.208 -0500 INFO Metrics - kb=90.31738
04-21-2020 23:46:48.203 -0500 INFO Metrics - kb=3.498046
04-21-2020 23:43:42.206 -0500 INFO Metrics - kb=73.24316
```

The preceding events show the number of kilobytes ingested in a Splunk indexer over a period of time. The field *kb* stores this numerical value. If you want to find the total data ingested during this period of time, you would need to sum the values of kb from all the three events. Using stats, you would simply run the following SPL:

```
...| stats sum(kb)
```

Splunk will produce a single value as a result:

```
sum(kb)
-------
167.058586
```

Note The preceding example does not show the SPL before the pipe (|) as it is not the significant part of the query. Many examples in this book follow this pattern.

Syntax

At its core, `stats` command utilizes a statistical function over one or more fields, and optionally splitting the results by one or more fields. The basic syntax of stats is shown in the following:

```
stats stats-function(field) [BY field-list]
```

As you can see, you must provide a *stats-function* that operates on a *field*. In our previous example, sum is the stats-function and kb is the field. The BY clause is optional. Unless the BY clause is specified, stats command produces one row as the output. If BY clause is specified with a field list, the result is split by the distinct field values. The by clause acts similar to the group by clause in SQL (Structured Query Language). You can calculate multiple values by separating them with comma or space as shown in the following:

```
stats stats-function(field),stats-function(field)…
```

In the following example, you calculate the minimum, maximum, and average kilobytes indexed:

```
…| stats min(kb),max(kb),avg(kb)
```

Splunk produces the following result:

```
min(kb)     max(kb)      avg(kb)
-------------------------------------
3.498046    90.317380    55.68619533333333
```

A very useful and common clause in SPL is AS, which is used to rename the fields for meaningful reporting. For example:

```
…| stats min(kb) AS "Minimum KB", max(kb) AS "Maximum KB", avg(kb) AS "Average KB"
```

And Splunk produces the following result:

```
Minimum KB   Maximum KB    Average KB
------------------------------------------------
3.498046     90.317380     55.68619533333333
```

Caution If the new field name has spaces in them, you must enclose the field name with double quotes. Exception to this is where and eval commands. When using these commands, surround them using single quotes. where and eval are covered later in this book.

Most users are unaware that you can also use wildcards in field names. Consider the following events:

```
04-21-2020 23:47:19.208 -0500 INFO Process - app_Time=56.56 db_Time=3.45 web_Time=10.34
04-21-2020 23:47:19.208 -0500 INFO Process - app_Time=123.75 db_Time=10.59 web_Time=45.76
04-21-2020 23:47:19.208 -0500 INFO Process - app_Time=87.99 db_Time=68.01 web_Time=19.18
```

There are three categories of response times found in these events: *app_Time, db_Time,* and *web_Time*. You can calculate average response times of all the three categories using the following SPL:

```
| stats avg(*_Time)
```

And Splunk produces the following output:

```
avg(app_Time)       avg(db_Time)        avg(web_Time)
-----------------------------------------------------------
89.43333333333334   27.350000000000005  25.093333333333334
```

If you want to rename the field names, you must specify wildcard in the AS clause as well:

```
...| stats avg(*_Time) AS "* Time"
```

And Splunk produces the following output:

```
app Time            db Time             web Time
-----------------------------------------------------------
89.43333333333334   27.350000000000005  25.093333333333334
```

If you don't specify any field in the stats-function, it implies all fields. So, avg() is equivalent to avg(*).

Counting Events

The simplest yet very useful application of `stats` is counting. The most basic form of this command is shown here:

```
...| stats count
```

The preceding command produces the total number of events in the result set. For example, consider the following events:

```
04-21-2020 23:47:19.208 -0500 INFO Metrics - kb=90.31738
04-21-2020 23:46:48.203 -0500 INFO Metrics - kb=3.498046
04-21-2020 23:43:42.206 -0500 INFO Metrics - kb=73.24316
```

For the SPL query ...| stats count, Splunk produces the following result:

```
Count
-----
3
```

Counting a Field

You can also specify a particular field as a parameter to the count function. For example, the following command will only count the events where the field *kb* is not empty:

```
...| stats count(kb)
```

Splunk produces the following result:

```
count(kb)
---------
3
```

Counting Using eval Expressions

You can also use eval to evaluate an expression for counting. eval is an SPL command that evaluates an expression and stores the result in a field. For example, consider the following events: to calculate the number of events where database response time was higher than ten milliseconds:

```
04-21-2020 23:47:19.208 -0500 INFO Process - app_Time=56.56 db_Time=3.45 web_Time=10.34
04-21-2020 23:47:19.208 -0500 INFO Process - app_Time=123.75 db_Time=10.59 web_Time=45.76
04-21-2020 23:47:19.208 -0500 INFO Process - app_Time=87.99 db_Time=68.01 web_Time=19.18
```

To calculate the number of events where database response time was higher than 10 milliseconds:

```
...| stats count(eval(db_Time > 10)) AS slow_db_trans
```

Splunk produces the following result:

```
slow_db_trans
-------------
2
```

Two events with *db_Time* values greater than 10 were found (10.59 and 68.01) and the `stats` command correctly retuned 2.

Caution When you use `eval` expression to count events, the resulting field must be renamed using the AS clause.

Eval is discussed in detail later in this chapter.

You can add as many stats-functions as you want separated by comma. For example:

```
...| stats
count(eval(db_Time <= 5)) AS "Normal",
count(eval(db_Time > 5 AND db_Time <= 10)) AS "Slow",
count(eval(db_Time > 10)) AS "Very slow"
```

Splunk produces the following result:

```
Normal   Slow    Very slow
---------------------------
1        0       2
```

Calculating Distinct Count

In some data sets, you may want to calculate the number of unique values of a field. You can do that using `distinct_count` function. Using the Splunk tutorial data

```
index=main sourcetype=access_combined_wcookie
| stats distinct_count(clientip) AS "Unique Client IPs"
```

Splunk produces the following result:

```
Unique Client IPs
-----------------
182
```

Note that `distinct_count` is very memory intensive, so you will need to be careful in running this function against large data sets. Splunk provides an alternative function named `estdc` which can be used to calculate estimated unique values:

```
index=main sourcetype=access_combined_wcookie
| stats estdc(clientip) AS "Unique Client IPs"
```

Splunk produces the following result:

```
Unique Client IPs
-----------------
182
```

Splitting the Results

Spitting the results by a field or list of fields is one of the most powerful features of SPL. It is achieved using the BY clause in SPL. In practice, you will most likely use the BY clause than any other clauses provided by SPL. For example, in order to calculate the events by HTTP method in the Splunk tutorial data

```
index=main sourcetype=access_combined_wcookie
| stats count BY method
```

Splunk produces the following result:

```
method    count
---------------
GET       24866
POST      14666
```

The field in the BY clause is known as *row-split* field as it splits the rows by the given field. You can add more than one row-split field to produce granular statistics. For example, to split the rows by method and action

```
index=main sourcetype=access_combined_wcookie
| stats count BY method, action
```

CHAPTER 2 CALCULATING STATISTICS

Splunk produces the following result:

```
method      action            count
-------------------------------------
GET         addtocart         2260
GET         changequantity    1105
GET         purchase           188
GET         remove            1173
GET         view              4358
POST        addtocart         3483
POST        changequantity    297
POST        purchase          5549
POST        remove             272
POST        view              1033
```

To split the results even further

```
index=main sourcetype=access_combined_wcookie
| stats count by method, action, categoryId
```

Splunk produces the following result:

```
method      action            categoryId       count
-----------------------------------------------------
GET         addtocart         ACCESSORIES      70
GET         addtocart         ARCADE           85
GET         addtocart         NULL             87
GET         addtocart         SHOOTER          46
GET         addtocart         SIMULATION       37
...
...
POST        addtocart         TEE              18
POST        changequantity    ACCESSORIES       5
POST        changequantity    ARCADE           17
POST        changequantity    NULL              9
...
...
```

46

Note that while wildcards are allowed in the field names in the stats-function, they are not allowed in the field names used in the BY clause. For example, the following SPL will not produce any results:

```
index=main sourcetype=access_combined_wcookie
| stats count BY "referer*"
```

Even though there are two fields named `referer` and `referer_domain` in the data, Splunk will not produce any results for the preceding query because of the wildcard in the BY clause.

Tip The keyword BY is not case sensitive.

Producing Aggregation Statistics

The `stats` command provides many aggregate functions such as average, sum, and percentiles. They can be extremely useful in creating useful reports, visualizations, and alerts. For example, to produce a report about the web requests in Splunk tutorial data

```
index=main sourcetype=access_combined_wcookie
| stats
max(bytes) AS "Largest",
min(bytes) AS "Smallest",
avg(bytes) AS "Average",
perc95(bytes) AS "95th percentile"
```

Splunk produces the following result:

Largest	Smallest	Average	95th percentile
4000	200	2097.7292067186077	3810.160296964131

Note 95th percentile indicates that 95% of the values are below this value. It can be a useful metric as averages tend to be affected by the outliers.

CHAPTER 2 CALCULATING STATISTICS

As another example, let us produce a useful report about the app response time from the following events:

```
04-21-2020 23:47:19.208 -0500 INFO Process - app_Time=56.56 db_Time=3.45 web_Time=10.34
04-21-2020 23:47:19.208 -0500 INFO Process - app_Time=123.75 db_Time=10.59 web_Time=45.76
04-21-2020 23:47:19.208 -0500 INFO Process - app_Time=87.99 db_Time=68.01 web_Time=19.18
...| stats median(app_Time) as "Median", range(app_Time) AS "Range"
```

Splunk produces the following result:

```
Median   Range
---------------
87.99    67.19
```

If you are interested in showing useful aggregations for a field, there is a useful command in SPL called fieldsummary. For example:

```
...| fieldsummary app_Time
```

Splunk produces the following result:

```
field     count   distinct_count   is_exact   max       mean                 min     numeric_count   stdev   values
-------------------------------------------------------------------------------------------------------------------
app_Time  3       3                1          123.75    89.43333333333334    56.56   3
33.618245542165525   [{"value":"123.75","count":1},{"value":"56.56","count":1},{"value":"87.99","count":1}]
```

Listing Unique Values

A useful feature of stats is to list the unique values of a field. For example, if you want to list the unique URI's accessed which resulted in 500 series HTTP errors (which indicates a server-side error)

```
index=main sourcetype="access_combined_wcookie" status=5*
| stats
count AS "Total Server Errors",
```
values(uri_path) AS "Unique URI accessed"
```
BY status
```

Splunk produces the following result:

```
status     Total Server Errors     Unique URI accessed
-------------------------------------------------------
500        733                     /cart.do
                                   /category.screen
                                   /oldlink
                                   /product.screen
503        952                     /cart.do
                                   /category.screen
                                   /oldlink
                                   /product.screen
505        480                     /cart.do
                                   /category.screen
                                   /oldlink
                                   /product.screen
```

You can also use the function list which shows *all* the values of a field, compared to values function which shows the *unique* values. For example, if you want to list all the client IP addresses which received 400 series HTTP errors (which indicate a client-side error) over a particular period of time

```
index=main sourcetype="access_combined_wcookie" status=4*
| stats list(clientip)
```

An excerpt of the results produced by Splunk is shown in the following:

```
list(clientip)
--------------
60.18.93.11
69.175.97.11
69.175.97.11
```

69.175.97.11
203.223.0.20
203.223.0.20
74.125.19.106

Using Time-Based Functions

With stats, there are many time-based functions that you can use to get insights into your machine data. For example, to find the first time a HTTP status code 503 was received, which indicates *Service unavailable*

index=main sourcetype="access_combined_wcookie" status=503
| **stats earliest(_raw) AS "First event with HTTP status 503"**

Splunk produces the following result:

```
First event with HTTP status 503
--------------------------------
112.111.162.4 - - [10/May/2019:18:26:38] "GET /oldlink?itemId=EST-7&JSESSIONID=SD7SL8FF5ADFF4964 HTTP 1.1" 503 1207 "http://www.buttercupgames.com/category.screen?categoryId=NULL" "Mozilla/5.0 (Windows NT 6.1; WOW64) AppleWebKit/536.5 (KHTML, like Gecko) Chrome/19.0.1084.52 Safari/536.5" 704
```

I used the function earliest and the field *_raw* to show the full text from the event. You can use any field that you are interested in.

As you may expect, counter to earliest, there is latest function to find the newest event. To find the last event from a certain IP address

index=main sourcetype="access_combined_wcookie" clientip="117.21.246.164"
| **stats latest(_raw) AS "Last event from 117.21.246.164"**

Splunk produces the following result:

```
Last event from 117.21.246.164
------------------------------
117.21.246.164 - - [10/May/2019:18:36:08] "GET /category.screen?categoryId=ARCADE&JSESSIONID=SD9SL6FF8ADFF5015 HTTP 1.1" 200 2078 "http://www.buttercupgames.com/oldlink?itemId=EST-15" "Googlebot/2.1 (http://www.googlebot.com/bot.html)" 667
```

> **Tip** Many new Splunk users use `sort` command to find the earliest and latest events, which is very ineffective. You could simply use the `earliest` and `latest` functions with stats command.

The functions `earliest_time` and `latest_time` retrieve the earliest and latest times of the event. For example, to find the time of the earliest client error (HTTP status code in 400 series)

```
index=main sourcetype="access_combined_wcookie" status=4*
| stats earliest_time(_raw) AS "Time of first client error"
```

Splunk produces the following result:

```
Time of first client error
--------------------------------
1575505670.000000
```

As you can see, the time is in Unix format. Later in Chapter 7, "*Advanced SPL Commands*," we'll see a way to convert the Unix time into human-readable format using the `convert` command.

Another useful time-based function is `rate` which calculates the per-second rate of change. This function works on accumulating counter metric. For example, to find the rate of change of total kilobytes indexed on a Splunk indexer

```
index=_internal sourcetype=splunkd component=Metrics
| stats rate(total_k_processed)
```

Splunk produces the following result:

```
rate(total_k_processed)
-----------------------
2.7142044132318617
```

The formula used for calculating the rate of change is as follows:

```
(latest(total_k_processed) - earliest(total_k_processed)) / (latest_
time(total_k_processed) - earliest_time(total_k_processed))
```

CHAPTER 2 CALCULATING STATISTICS

Event Order Functions

There are two event order functions that you can use with `stats`, namely, `first` and `last`. The function `first` indicates the first seen value, which is usually the most recent instance of this field. The function `last` indicates the last seen value, which is usually the oldest instance of this field.

For example, to show the first and last values of *uri_path* accessed in the Splunk tutorial data

```
index=main sourcetype="access_combined_wcookie"
| stats first(uri_path) AS "Newest URI", last(uri_path) AS "Oldest URI"
```

Splunk produces the following result:

```
Newest URI      Oldest URI
------------------------------
/oldlink        /product.screen
```

> **Caution** `first` and `last` functions return events as seen by the stats command. They won't necessarily be chronological in order.

Now that you have a good understanding of `stats` and their various aggregation functions, let's move on to the cousins of `stats` – `eventstats` and `streamstats`.

Eventstats and Streamstats

While `stats` *command helps us with generating statistics from your machine data, their cousins* `eventstats` *and* `streamstats` *add even more features. All functions of* `stats` *command can be used with these two commands as well. We will explore* `eventstats` *first, followed by* `streamstats`.

Eventstats

The command `eventstats` calculates statistics the same way `stats` does, except it saves the statistics as a field in *all* the resulting events. With `stats`, the result is just a table of statistics. With `eventstats`, the resulting statistics is added to *all* the events. This means you can not only use the statistics in the subsequent commands, you can also use all the

CHAPTER 2   CALCULATING STATISTICS

original fields. You can use all the aggregation function that are available with the stats command. Let us see this with an example. Consider the following events:

```
04-21-2020 23:47:19.208 -0500 INFO Metrics - kb=90.31738
04-21-2020 23:46:48.203 -0500 INFO Metrics - kb=3.498046
04-21-2020 23:43:42.206 -0500 INFO Metrics - kb=73.24316
```

Let us simply print the values in a table using the table command. The table command simply takes a list of fields as parameters and prints them in a table:

```
...| table _time,kb
```

Splunk produces the following result:

```
_time                       kb
--------------------------------
2020-04-21 23:46:48.203    3.498046
2020-04-21 23:43:42.206    73.24316
2020-04-21 23:47:19.208    90.31738
```

Now, let's use the eventstats command to calculate the average kb and add the result to each event:

```
...| eventstats avg(kb) AS "Average KB"
| table _time,kb,"Average KB"
```

Splunk produces the following result:

```
_time                       kb          Average KB
----------------------------------------------------
2020-04-21 23:46:48.203    3.498046    55.68619533333333
2020-04-21 23:47:19.208    90.31738    55.68619533333333
2020-04-21 23:43:42.206    73.24316    55.68619533333333
```

Notice how the *Average KB* is added to each event. One of the primary applications of eventstats is to use the statistics in visualizations. For example, you can plot a metric such as *response time* over a period of time, and also show the *average response time* in the same graph. This helps to visually inspect if a metric is over the average at any time.

53

CHAPTER 2 CALCULATING STATISTICS

Streamstats

The command `streamstats` calculates cumulative statistics of a field as the events are seen in a streaming manner. For example, you can keep a running count of a metric on a continuous basis. You can use all the aggregation functions that are available with the `stats` command. Consider the same set of events that we used in our previous example:

```
04-21-2020 23:47:19.208 -0500 INFO Metrics - kb=90.31738
04-21-2020 23:46:48.203 -0500 INFO Metrics - kb=3.498046
04-21-2020 23:43:42.206 -0500 INFO Metrics - kb=73.24316
```

If you want to keep a running sum of the *kb* field, you can use `streamstats` as shown in the following:

```
...| streamstats sum(kb) AS total_kb
| table _time,kb,total_kb
```

Splunk produces the following result:

```
_time                     kb          total_kb
----------------------------------------------
2020-04-21 23:47:19.208   90.31738    90.31738
2020-04-21 23:46:48.203   3.498046    93.815426
2020-04-21 23:43:42.206   73.24316    167.058586
```

One thing you notice is the order of the events. By default, Splunk sees the **latest event first**. To calculate the running count of a metric, it is preferred to count from the earliest event. In order to do this, you reverse the events first with the `reverse` command:

```
...| reverse
| streamstats sum(kb) AS total_kb
| table _time,kb,total_kb
```

Splunk produces the following result:

```
_time                     kb          total_kb
----------------------------------------------
2020-04-21 23:43:42.206   73.24316    73.24316
2020-04-21 23:46:48.203   3.498046    76.741206
2020-04-21 23:47:19.208   90.31738    167.058586
```

Another classic example of using `streamstats` is *ranking*. Using the Splunk tutorial data, if you want to rank the top five *itemIds*

```
index=main sourcetype=access_combined_wcookie action=purchase
| stats count AS "total_purchases" BY itemId
| sort 5 - total_purchases
| streamstats count AS rank
```

Splunk produces the following result:

```
itemId     total_purchases     rank
-------------------------------------
EST-15     449                 1
EST-14     439                 2
EST-21     438                 3
EST-26     415                 4
EST-7      414                 5
```

First, we filter the events using *action=purchase*; then, we utilize the `stats` command to find the count of events by *itemId*. We then sort them to find the top five *itemIds*, and finally we use the `streamstats` command to add the count field, which simply increments as a new event is seen.

A useful option of `streamstats` command is `window`. Using this option, you can specify the number of events to use when calculating statistics. For example, to calculate the average of the field *response_time* over the last ten events

```
...| streamstats avg(response_time) window=10
```

The preceding SPL query will calculate the average response time in a streaming manner for up to ten events. After ten events, the value is reset and starts over again for the next set of ten events.

In addition to `stats`, `eventstats`, and `streamstats`, there are two additional `stats` family commands that you can use. They are `tstats` and `mstats`. The command `tstats` is used to produce statistics out of indexed fields and accelerated data models. The command `mstats` operates on metric data store. We will cover `tstats` in detail later in this book in Chapter 7, "Advanced SPL Commands." We will not discuss `mstats` as it is out of the scope of this book. You can obtain documentation from docs.splunk.com if you want to learn about `mstats`.

CHAPTER 2 CALCULATING STATISTICS

Two handy commands for identifying the most frequent and least frequent value of a field are top and rare. We'll take a look at them next.

Using Top and Rare

Using top and rare commands, you can find the most common values and least common values of a field, respectively (i.e., the frequency of the field values). By default, both top and rare show ten values. You can override this by specifying the number of values to retrieve. For example, to find the top five vendor id from the Splunk tutorial data

```
index=main sourcetype=vendor_sales
| top limit=5 VendorID
```

Splunk produces the following result:

VendorID	count	percent
1060	135	0.446370
7014	133	0.439757
1015	128	0.423224
1005	128	0.423224
1010	126	0.416612

The keyword limit is optional. For example, the following command will produce the same result:

```
index=main sourcetype=vendor_sales
| top 5 VendorID
```

The top command also outputs the percent for each vendor ID. You can disable the percent field if you want, by using the showperc option:

```
index=main sourcetype=vendor_sales
| top limit=5 VendorID showperc=f
```

Splunk produces the following result:

```
VendorID   count
---------------
1060       135
7014       133
1015       128
1005       128
1010       126
```

> **Tip** In showperc=f, instead of f, you can also use false or 0. Similarly, you can use t or true or 1 where required.

In order to find the lowest selling product using the Splunk tutorial data

index=main sourcetype=access_combined_wcookie action=purchase
| **rare limit=5 productId showperc=f**

Splunk produces the following result:

```
productId      count
--------------------
SF-BVS-01      1
SF-BVS-G01     62
CU-PG-G06      148
BS-AG-G09      151
FI-AG-G08      163
```

You can specify more than one field to derive the top statistics. For example, if you want to list the top two HTTP errors for *all* the *actions* using the Splunk tutorial data

index=main sourcetype=access_combined_wcookie status != 200
| **top action,status limit=2 showperc=f**

CHAPTER 2   CALCULATING STATISTICS

Splunk produces the following result:

```
action      status    count
-------------------------------
purchase    503       298
view        500       145
```

You can also list the top two HTTP errors for *each action*; you can use the by clause:

```
index=main sourcetype=access_combined_wcookie status != 200
| top status by action limit=2 showperc=f
```

Splunk produces the following result:

```
action            status    count
-----------------------------------
addtocart         408       88
addtocart         500       72
changequantity    503       37
changequantity    400       36
purchase          503       298
purchase          408       44
remove            500       47
remove            503       33
view              500       145
view              408       136
```

You can either completely disable the count field or rename as you prefer. To disable, use `showcount=f`, and to rename, use `countfield=<New field name>`.

Next, we'll focus on another popular command for producing statistics and visualizations – `chart`.

Chart

Chart is another very useful transforming command that orders result in a table. Similar to `stats`, output of chart can be used to create visualizations such as column or pie charts. Consider the following sample events:

```
server=pluto  status=success  response_time=345   state=IL
server=earth  status=success  response_time=231   state=MN
server=earth  status=fail     response_time=5023  state=WI
server=venus  status=success  response_time=120   state=MN
server=pluto  status=success  response_time=65    state=MN
server=earth  status=success  response_time=438   state=IL
server=saturn status=fail     response_time=5045  state=MN
```

Let us look at an example:

```
...| chart count by server
```

Splunk produces the following result:

```
server    count
--------------
earth     3
pluto     2
saturn    1
venus     1
```

In this case, where we have one field in the by clause, chart behaves exactly like stats. But when we add a second field to the by clause, chart behaves differently:

```
...| chart count BY server, status
```

Splunk produces the following result:

```
server    fail    success
-------------------------
earth     1       2
pluto     0       2
saturn    1       0
venus     0       1
```

The first field in the by clause is the *row-split* field, and the second field is the *column-split* field. The *row-split* field becomes the first column, and the values become the row labels. When a *column-split* field is specified, each column represents a distinct value of the *column-split* field. In our example, the two distinct values of status are *fail* and *success*.

CHAPTER 2 CALCULATING STATISTICS

Consider the same example with `stats` command:

`...| stats count BY server, status`

Splunk produces the following result:

```
server     status      count
------------------------------
earth      fail        1
earth      success     2
pluto      success     2
saturn     fail        1
venus      success     1
```

As you can see, with stats, each row is a unique combination of the group by fields. This allows for granular grouping. For example:

`...| stats count BY server,status, state`

Splunk produces the following result:

```
server     status      state     count
----------------------------------------
earth      fail        WI        1
earth      success     IL        1
earth      success     MN        1
pluto      success     IL        1
pluto      success     MN        1
saturn     fail        MN        1
venus      success     MN        1
```

If you try to run the same command with `chart`, you will get an error:

`...| chart count BY server,status, state`

Splunk produces the following error:

`Error in 'chart' command: The argument 'state' is invalid.`

CHAPTER 2 CALCULATING STATISTICS

The chart command accepts a maximum of two split-by fields, one for *row-split* and other for *column-split*. On the other hand, stats command can accept any number of fields. In addition, when you produce chart visualizations, the *row-split* field in the chart command becomes the x axis.

Some users get confused to see the over keyword used with chart. Let me clarify. With chart command, you can use the over keyword to specify the *row-split* field followed by the by clause to specify the *column-split* field. For example, the following two queries are equivalent:

```
...| chart count over server BY status
```

```
...| chart count BY server,status
```

Both the preceding commands produce the same result:

```
server     fail     success
-----------------------
earth      1        2
pluto      0        2
saturn     1        0
venus      0        1
```

A useful option of chart command is span. Using span, you can break down the results by a meaningful range. For example, to break down the *bytes* field by 1000 from the Splunk tutorial data

```
index="main" sourcetype=access_combined_wcookie
| chart count BY bytes span=1000
```

Splunk produces the following result:

```
bytes        count
---------------
0-1000       8340
1000-2000    10384
2000-3000    10455
3000-4000    10339
4000-5000    14
```

This kind of table provides us a great understanding of the data spread. For example, you can readily see that the majority of the responses were between 1000 and 4000 bytes in size, with a very few responses above 4000 bytes.

You can use eval functions just like in stats. For example, to plot the number of failures by server in our sample events

```
...| chart count(eval(response_time > 5000)) AS "Slow transactions" BY server
```

Splunk produces the following result:

```
server    Slow transactions
----------------------------
earth     1
pluto     0
saturn    1
venus     0
```

The chart command is particularly useful in producing visualizations due to its column-split functionality. Next, we'll talk a look at yet another useful command in detail that can be used with many statistical commands, eval.

Eval

We have covered a lot of ground with statistical commands. The command eval can be considered as a ubiquitous command that can aid many SPL commands. It is especially powerful when used with statistical commands. To put it simply, eval command calculates an expression and stores the result into a field. If the field already exists, it overwrites the value of the field. When used within statistical commands, eval command dynamically creates the file. Note that eval command does not rewrite data into the index. Let us consider a basic example using the same sample data we used earlier:

```
04-21-2020 23:47:19.208 -0500 INFO Metrics - kb=90.31738
04-21-2020 23:46:48.203 -0500 INFO Metrics - kb=3.498046
04-21-2020 23:43:42.206 -0500 INFO Metrics - kb=73.24316
```

CHAPTER 2 CALCULATING STATISTICS

Consider the following eval command:

`...| `**`eval bytes = kb * 1024`**` | table kb,bytes`

Splunk produces the following result:

```
Kb              bytes
------------------
3.498046        3581.999
73.24316        75001.00
90.31738        92485.00
```

The preceding command takes the value of the field *kb* and multiples by 1024 and stores the result in a new field called *bytes* in every event where the field *kb* is present. We'll now take a look at the eval expressions in depth.

Eval Expressions

An eval expression can contain any of the following:

1. Functions
   Example: case,match,like

2. Arithmetic operators
   Example: +,-,*,/,%

3. String concatenation
   Example: Using . (period) to concatenate two string

4. Logical operators
   Example: AND, OR, NOT, XMOR, < , >, <=, >=,=,==,!=,LIKE

Let us learn how to use eval by looking at various applications.

Calculating

With eval, you can use many mathematical operators and functions to calculate values.
Calculate error percentage:

`...| eval error_percentage = number_of_errors/total_transactions * 100`

Calculate cosine of y and assign to x:

```
...| eval x = cos(y)
```

Calculate a SHA256 hash of the string *mypassword*:

```
...| eval hash=sha256("mypassword")
```

Calculate length of a string:

```
...| eval length = len("Apress")
```

The preceding SPL returns length=6.
Calculate X to the power of Y:

```
...| eval answer = pow(X,Y)
```

Calculate square root of a number:

```
...| eval root = sqrt(16)
```

The preceding SPL returns root=4.
Produce a random number between 0 and 1000:

```
...| eval random_num = random() % 1000
```

The `random` function produces a number between 0 and 2147483647. By using the modulo operator %, which returns the reminder of the random number divided by 1000, you can ensure that *random_num* will always be less than 1000.

Converting

You can convert numbers to strings and vice versa with eval.
Convert the number stored in the field `claimId` to string:

```
...| eval claimIdString = toString(claimId)
```

If the parameter to the `toString` function is Boolean, it results in either `True` or `False`. In addition, you can use the following optional parameters when converting a number to a string.

CHAPTER 2 CALCULATING STATISTICS

To convert a number to hexadecimal, you can use the hex option:

```
...| eval num_hex = toString(num,"hex")
```

The preceding SPL converts *num* to hexadecimal and assigns the result to the *num_hex* field.

To convert a number to a string with commas, you can use commas option:

```
...| eval num_with_commas = toString(num,"commas")
```

For example, the number 4589347294 will result in 4,589,347,294.

To convert the number of seconds to an easily recognizable format, you can use duration option:

```
...| eval Time = toString(seconds,"duration")
```

For example:

```
...| eval seconds = 100
| eval Time = toString(seconds,"duration")
| table seconds, Time
```

Splunk produces the following result:

```
seconds    Time
------------------
100        00:01:40
```

You can use printf function to format the value and store it in a field:

```
...| eval pi_value = printf("%.4f",pi())
```

The preceding SPL returns *pi_value* = 3.1416. The printf function is similar to languages like python and C.

You can convert a string to a number using toNumber function:

```
...| eval order_id_num = toNumber(order_id)
```

Eval also provides many ways to format data. Let us look at them in detail.

65

Formatting

You can use eval functions to format your results.

To transform a string to lowercase

```
...| eval lname = lower("Apress Book")
```

The preceding SPL returns lname="apress book".

As you may have guessed, to transform a string to uppercase

```
...| eval uname = upper("Apress Book")
```

The preceding SPL returns uname="APRESS BOOK".

To trim a string X with characters in Y from left

```
...| eval ltname = ltrim("Mr. James", "Mr. ")
```

The preceding SPL returns ltname="James".

To trim a string X with characters in Y from right:

```
...| eval rtname = rtrim("James Jr.", "Jr.")
```

The preceding SPL returns rtname="James".

To trim a string X with characters in Y from both the sides

```
...| eval tname = trim("zzzJameszz", "z")
```

The preceding SPL returns tname="James".

Note With the `trim` functions, if the second parameter is not provided, spaces and tabs are removed.

To decode a url-encoded string

```
...| eval url = urldecode("https%3A%2F%2Fwww.apress.com%2Fus")
```

The preceding SPL returns url = www.apress.com/us.

To concatenate two or more strings, you can use the . (period) operator:

```
...| eval profit = "$".(revenue - expense)
```

The preceding SPL calculates *revenue – expense* and adds the string "$" at beginning.

CHAPTER 2   CALCULATING STATISTICS

To format a Unix time as a string with a specific format

```
... | eval now = strftime(now(),"%I:%M:%S %p")
```

A sample output can be 07:24:55 PM. For a listing of common date and time formatting, see https://docs.splunk.com/Documentation/Splunk/latest/SearchReference/Commontimeformatvariables.

To format a time in string into a Unix time

```
...| eval mytime = strptime("05/06/2020 18:35:24","%d/%m/%Y %H:%M:%S")
```

The preceding SPL returns mytime= 1591400124.000000.

Eval provides a few useful commands to round numbers. Let's learn about them next.

Rounding

To round a number to two decimal points

```
...| eval rnum = round(5.3456,2)
```

The preceding SPL returns rnum=5.35.

To round a number to the next highest integer

```
...| eval cnum = ceil(5.3456)
```

The preceding SPL returns cnum=6.

To round a number down to the nearest integer

```
...| eval cnum = floor(5.3456)
```

The preceding SPL returns cnum=5.

One of the reasons `eval` is very powerful is due to the logical operations it can perform. We'll take a look at them now.

Performing Conditional Operations

You can use many conditional operations with eval. This is one of the most commonly used advanced techniques of using eval.

You can use the `if` function to evaluate a logic expression. For example:

```
...| eval type = if(num % 2 == 0,"Even","Odd")
```

67

The preceding SPL returns type=Even for even numbers and type=Odd for odd numbers. The `if` function's first parameter is a logical expression. If the expression evaluates to true, the second parameter is returned. If the expression evaluates to false, the third parameter is returned.

You can use the `like` function to check if a string matches a certain pattern. The like function returns true if the pattern matches and false if it doesn't match. See the following example:

```
...| eval errorType = if(like(error, "java%Exception"),"Java Exception","Unknown Error")
```

If the field error contains values such as *java.lang.NullPointerException, java.net.ConnectionException*, the *errorType* will be set to Java Exception. Otherwise, it will be set to Unknown Error. The character % is used as a wildcard to match multiple characters. The character _ (underscore) can be used to match one character.

For a full regular expression matching, you can use the `match` function:

```
... | eval SSN_found = if(match(_raw,"\d{3}-\d{2}-\d{4}"),"Yes","No")
```

The preceding SPL returns SSN_found = Yes if the raw event contains a social security number patter (xxx-xx-xxxx). Otherwise, it returns SSN_found = No.

You can use `case` function if you have multiple values to match. For example:

```
index=main sourcetype=access_combined_wcookie
| eval httpStatusCategory = case(status == 200, "OK", status >=400 AND status < 500, "Client Error", status >= 500, "Server Error", 1==1, "N/A")
| stats count by httpStatusCategory
```

Splunk produces the following result:

```
httpStatusCategory      count
------------------------------
Client Error            3085
OK                      34282
Server Error            2165
```

The `case` function accepts a series of logical expressions and defines the values to be returned based on the evaluations of those expressions. For example, the first logical expression `status == 200` checks if *status i*s equal to 200. If true, the string "OK" is returned. If false, processing proceeds to the next logical expression. If the *status* is between 400 and 500, denoted by the expression `status >=400 AND status < 500`, the string `"Client Error"` is returned. Finally, if the status is greater than or equal to 500, denoted by the expression `status >= 500`, the string `"Server Error"` is returned. As a *catch-all-other* expression, if none of the previous logical expressions evaluated to be true, the string `"N/A"` is returned. Using `"1==1"` always results in true and is a usual practice to use this as the *catch-all-other* expression.

We've covered `eval` in detail. Combining `eval` and `stats` is one of the most common ways to produce meaningful reports. In addition to searching and producing statistics, Splunk excels at creating visualizations of data. In the next section, let us briefly look at how you can create stunning visualizations almost effortlessly.

Creating Visualizations

With Splunk, creating visualizations is instantaneous, in the sense that the results of your statistical commands automatically create the visualization that you can customize to your needs. There are varieties of visualizations such as column chart, area chart, pie chart, single value gauges, and choropleth graphs that you can create using Splunk. While this book does not discuss visualizations in depth, a quick overview is provided so that you can put them to use right away.

Not all searches can be rendered as visualizations. In order to create visualizations, your SPL query must produce some statistics. This means you have to utilize statistical commands such as `stats`, `top,` and `chart`. The resulting visualization can be saved as a report that can be run anytime, or as a dashboard panel. You can even schedule your reports and dashboards to run at certain times and email the results as PDFs.

For example, to create a visualization of the top five *categoryId* using Splunk tutorial data, first execute the following SPL query:

```
index=main sourcetype=access_combined_wcookie action=purchase
| top limit=5 categoryId showperc=f
```

CHAPTER 2 CALCULATING STATISTICS

Splunk produces the following result:

```
categoryId          count
--------------------------
STRATEGY            885
ARCADE              537
TEE                 404
ACCESSORIES         387
SHOOTER             275
```

Splunk automatically takes you to the Statistics tab as shown in Figure 2-1.

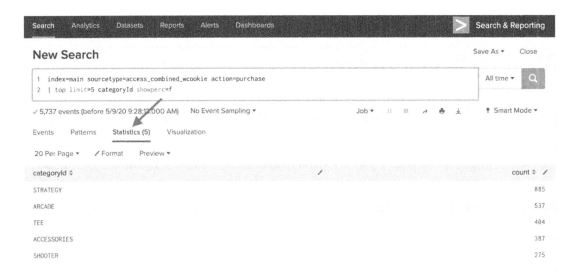

Figure 2-1. *For statistical searches, Splunk automatically takes you to the Statistics tab*

Click on the Visualizations tab to see the visualization that has been created for you. See Figure 2-2.

CHAPTER 2 CALCULATING STATISTICS

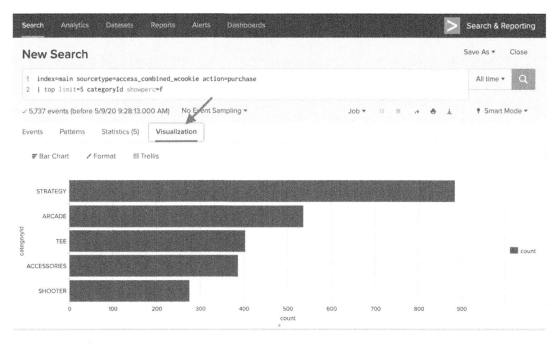

Figure 2-2. *Visualization tab with the automatically generated visualization*

The SPL commands that are generally used to create visualizations

1. stats
2. chart
3. top
4. rare
5. timechart

We will use examples from the first four commands in this chapter. We'll learn about `timechart` command and associated visualizations in the next chapter.

Note While the statistical commands are the most common way to create visualizations, any search results that are formatted into a table can be used to create visualizations.

CHAPTER 2   CALCULATING STATISTICS

Switching the Type of Visualization

Once a visualization is generated, it is easy to switch to various types of visualizations. This is helpful as you can try out various options before saving your visualization as a report or a dashboard. In order to switch visualizations, simply click on *Select Visualization* button, which is labeled based on the current active visualization. See Figure 2-3.

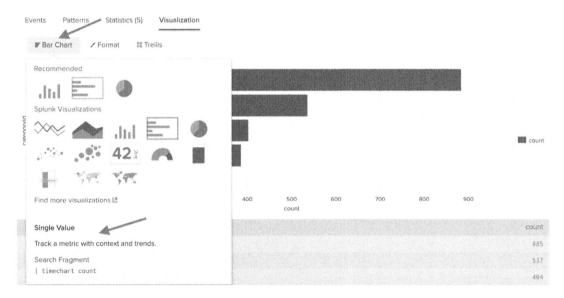

Figure 2-3. *Switching visualizations*

As you hover over each visualization, Splunk provides a short description of the visualization and typical search command used to create that visualization. Let us learn more about at the most common types of visualizations.

Line Chart

A line chart plots the data with values represented as a line. This is one of the basic forms of visualization. For example:

```
index=main sourcetype=access_combined_wcookie action=purchase
| stats count by status
```

CHAPTER 2 CALCULATING STATISTICS

Splunk produces the following result:

```
Status       count
---------------
200          5224
400          37
403          9
404          21
406          37
408          44
500          34
503          298
505          33
```

A line chart visualization produces the chart shown in Figure 2-4.

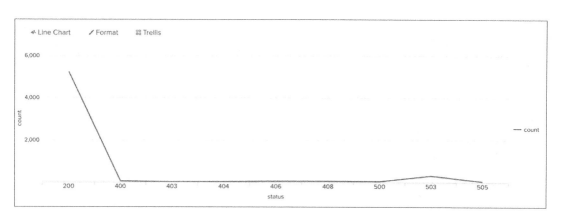

Figure 2-4. *A line chart*

Area Chart

An area chart is very similar to line chart, except the area under the line is shaded. This can be useful to make a contrasting difference in the rendering of the visualization. See Figure 2-5:

```
index=main sourcetype=access_combined_wcookie action=purchase
| stats count by status
```

CHAPTER 2 CALCULATING STATISTICS

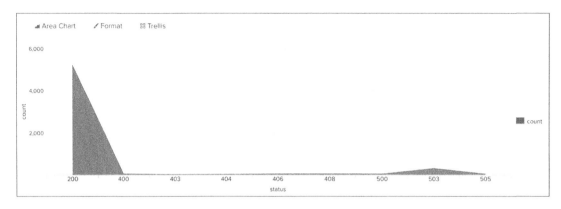

Figure 2-5. *An area chart*

Column Chart

A column chart plots the data as bars stacked as columns. This form of visualization is useful for comparing data. See Figure 2-6:

```
index=main sourcetype=access_combined_wcookie action=purchase
| top limit=5 categoryId showperc=f
```

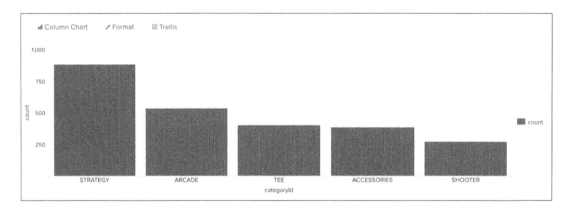

Figure 2-6. *A column chart*

Bar Chart

A bar chart, similar to column start, is suitable for comparing data. It represents data in horizontal bars. See Figure 2-7:

```
index=main sourcetype=access_combined_wcookie action=purchase
| top limit=5 categoryId showperc=f
```

CHAPTER 2 CALCULATING STATISTICS

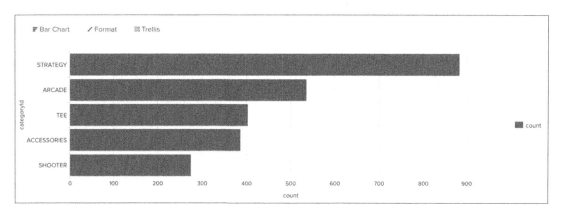

Figure 2-7. A bar chart

Pie Chart

A pie chart organizes data in a circle with slices of data representing unique values. It provides a great way for instant comparisons of data. See Figure 2-8:

```
index=main sourcetype=access_combined_wcookie action=purchase
| top limit=5 categoryId showperc=f
```

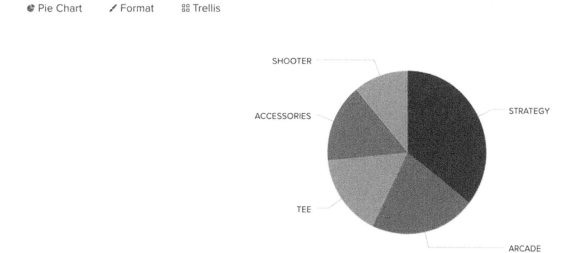

Figure 2-8. A pie chart

CHAPTER 2 CALCULATING STATISTICS

Plotting Multiple Data Series

The first column in the table resulting from a search becomes the x axis of a chart. The subsequent columns can be plotted as multiple data series. In most cases, `chart` command is more appropriate for producing multiple data series than `stats` command. This is because when you split the column using the *column-split* by clause of `chart` command, each column in the output table represents a distinct value of the *column-split* field. Each of these columns becomes a data series. When you use `stats` command however, each row in the output table is a unique combination of values of the split-by fields. They don't fit well for a chart. For example:

```
index=main sourcetype=access_combined_wcookie action=purchase
| chart count by categoryId,itemId useother=f limit=4
```

Splunk produces the following result:

categoryId	EST-14	EST-15	EST-21	EST-7
ACCESSORIES	34	30	25	29
ARCADE	29	35	44	33
NULL	11	4	5	6
SHOOTER	20	26	23	16
SIMULATION	27	24	16	22
SPORTS	10	9	14	9
STRATEGY	75	75	71	65
TEE	22	31	29	35

The SPL generates a data table in which the rows are the categories, denoted by the row-split field *categoryId*. Each column represents a unique value of the column-split field *itemId*. The parameter `limit=4` limits the number of columns to four based on the sum of the values in a column. By default, the chart limits the number of columns to ten, and the rest is aggregated in a column named OTHER. Using `useother=f` prevents this behavior.

A column chart visualization is shown in Figure 2-9.

CHAPTER 2 CALCULATING STATISTICS

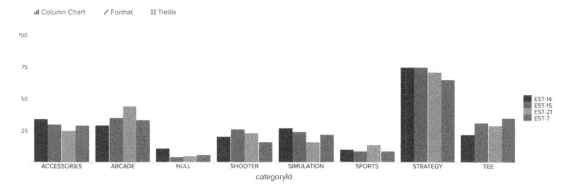

Figure 2-9. Plotting multiple data series in a chart

A useful way of visualizing multiple data series is using *stacked column chart*. To create a stacked column chart, click on the *Format* menu and choose *stacked* under *General* tab. See Figure 2-10.

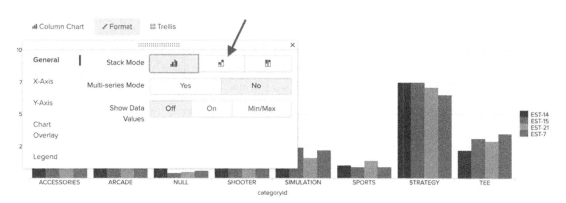

Figure 2-10. Formatting column chart with stacked columns

CHAPTER 2 CALCULATING STATISTICS

The resulting visualization looks like Figure 2-11.

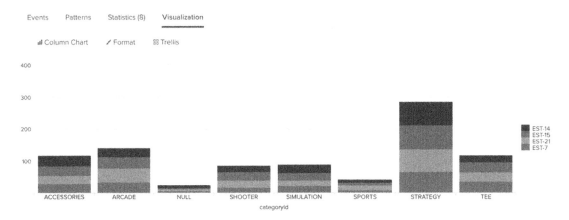

Figure 2-11. A stacked column chart

We have covered a lot of ground. Let us review the key takeaways from this chapter.

Key Takeaways

In this chapter, you learned one of the most important applications of SPL, generating statistics. We looked at the powerful `stats` command and many of its useful aggregation functions. We reviewed many examples to learn the various statistics you can create, such as *average, sum, percentile, max, min,* and *list*. In addition, we looked at two useful commands, `top` and `rare`. Further, we learned how to use `eventstats` and `streamstats`. We reviewed the `chart` command and how it can group the data using *row-split* and *column-split* fields. We then learned the `eval` command in detail to create and manipulate fields. Finally, we looked at how to use the statistics you generate to create useful visualizations. Here are the key takeaways from this chapter:

1. The `stats` command calculates statistics from the search results. It produces one row as result unless the BY clause is specified to split the result by a list of fields.

2. With `stats` command you can use 15+ aggregate functions such as `avg, sum, median,` and `percentile.`

3. Stats command also provides many time-based functions such as `earliest, latest,` and `rate.`

CHAPTER 2   CALCULATING STATISTICS

4. The command `eventstats` makes the statistics available in each event as a new field, which makes it possible to use all your fields in the events in subsequent commands.

5. You can use the command `streamstats` to calculate cumulative statistics in a streaming manner as each event is seen in the result.

6. `top` and `rare` commands are used to find the most common and least common values of the fields.

7. You can use the `chart` command to calculate statistics that are useful for creating visualizations. The first parameter of `chart` command is row-split field (x axis), and the second parameter is used as column-split field (y axis).

8. You can use the `eval` command to calculate expressions and place the result in a field that can be used in subsequent commands. If the field already exists, value is overwritten.

9. There are 80+ `eval` functions that can be used to calculate mathematical, date/time, comparison, and text expressions.

10. Results of statistical commands automatically create visualizations that can be viewed and customized in *Visualizations* tab.

By learning how to use the `stats` family of commands and `eval`, you are now equipped to create insightful statistics from your machine data through the power of SPL. In the next chapter, we'll learn another important application of SPL – using time and time-related operations. We will cover yet another important command in SPL repertoire – `timechart`.

CHAPTER 3

Using Time-Related Operations

Splunk is optimized for indexing time-series data. In this context, one could say that Splunk is a time-series database, even though it does much more than that. When Splunk indexes an event, it expects to retrieve the timestamp from the event and store it as a field along with the raw data. Even if the raw event does not contain the timestamp, Splunk will assign one at index time using various approximations, such as the time at which the event was indexed, the last modification time of the source file, the timestamp of the previous event, and so on. SPL provides a rich set of time-related commands to search and create visualizations over time. In this chapter, we'll learn how to use the most important time-specific SPL commands and functions.

Splunk and Time

Under the covers, Splunk stores data in *data buckets* which are directories with a specific structure residing in file system. Splunk names its data buckets (where raw data and index files are stored) using the timestamps. In this way, when you search Splunk, if a data bucket is not within the time range you specify in search, Splunk doesn't bother opening the bucket, thereby increasing the efficiency of the search. For example, consider the following data bucket name:

```
db_1575511972_1572226601_6
```

The number 1575511972 indicates the timestamp of the newest event in the bucket in epoch format, which translates to *Thursday, December 5, 2019 2:12:52 AM GMT*. The number 1572226601 denotes the oldest event in the bucket in epoch format, which translates to *Monday, October 28, 2019 1:36:41 AM GMT*.

CHAPTER 3   USING TIME-RELATED OPERATIONS

Note The hot buckets, which contain the newest data, do not reflect the time range in the bucket name. It is the warm and cold buckets that are named after the time range.

One of the ways you can configure data retention in Splunk is using the age of the data (you can also use the size of the index as a constraint). When data retention is configured using the age of the data, a data bucket expires when the earliest event in the bucket exceeds this time (this is configured using `frozenTimePeriodInSecs` in *indexes. conf*).

The timestamp of an event is always stored in Unix time (also known as epoch time) in the indexes. It is stored in the _time field. Storing the time in Unix time format ensures that your search results will be consistent regardless of the time zone you are in. Figure 3-1 highlights where time is displayed in the search interface.

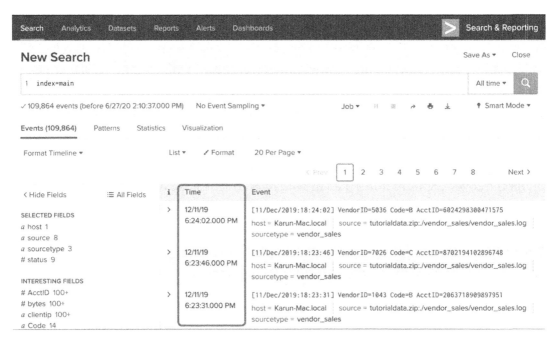

Figure 3-1. Time in search interface

Note that even though the time is stored in Unix time in the indexes, when displaying time in the search interface, Splunk displays the time in human-readable form.

CHAPTER 3 USING TIME-RELATED OPERATIONS

A Note About Time Zone

You can set the time zone of your local time using the Preferences menu under your username in Splunk Web. See Figure 3-2.

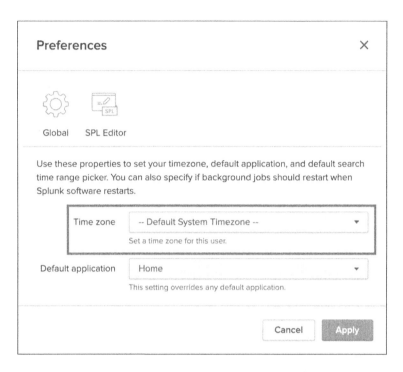

Figure 3-2. *Configuring time zone in Splunk Web*

When displaying time in the search interface, Splunk will automatically render the event time in your time zone.

Note that when you use relative time ranges in the time picker, different time zones may show different results. For example, when a user in Minneapolis (Central Time Zone) searches using *Today* as the time range, events since 12:00 AM central time will be retrieved. The same search will retrieve different results when a user from Chennai (Indian standard time) performs the same search. When you use time ranges such as *Last 24 hours,* Splunk uses Unix time to process the search, and the retrieved events are identical between different time zones.

Now that you have some background on Splunk and time, let's dive into the various SPL commands that help with searching and visualizing your data, starting with the most important of all, `timechart`.

CHAPTER 3   USING TIME-RELATED OPERATIONS

Timechart

In the previous chapter, we learned `stats`, `chart,` and `eval`. In this section, we'll learn `timechart`, another very useful command in the SPL repertoire. At a high level, `timechart` is very similar to the `chart` command, except that `timechart` always plots data with *time* on the x axis. You can optionally specify one by clause field. Each value of the by clause field becomes a series in the chart. Let's look at a quick example.

Using the Splunk tutorial data

```
index=main sourcetype=access_combined_wcookie
| timechart count
```

Set the time range to *All time* to ensure results are returned. Splunk automatically takes you to the *Statistics* tab and shows the following results:

_time	count
2019-12-04	1220
2019-12-05	5784
2019-12-06	5868
2019-12-07	5801
2019-12-08	5383
2019-12-09	5545
2019-12-10	5538
2019-12-11	4393

When you switch to the Visualization tab, you can see a graphical representation. See Figure 3-3.

CHAPTER 3   USING TIME-RELATED OPERATIONS

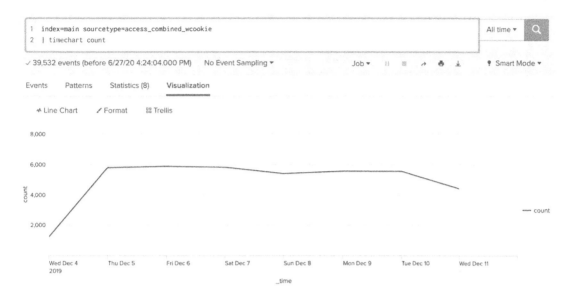

Figure 3-3. *A line chart created using the timechart command*

> **Note** You may have to select *line chart* from the list of visualizations.

This type of visualization is extremely useful during troubleshooting and analyzing your logs. For example, if there is a large spike or dip in the chart, that can indicate a significant event in your application.

Specifying Time Span

In the previous example, you may have noticed that Splunk automatically applied a scale to the x axis, which was 1 day. That is, each datum represents the measurement for 1 day. This is called the *span*. Splunk determines the span based on the time range of the search as shown in Table 3-1.

CHAPTER 3   USING TIME-RELATED OPERATIONS

Table 3-1. *Default time spans chosen by timechart based on the search-time range*

Time range	Default time span
Last 15 minutes	10 seconds
Last 60 minutes	1 minute
Last 4 hours	5 minutes
Last 24 hours	30 minutes
Last 7 days	1 day
Last 30 days	1 day
Previous year	1 month

This time span can be explicitly specified using the span option. Let us rewrite the previous search with a time span of 12 hours:

```
index=main sourcetype="access_combined_wcookie"
| timechart span=12h count
```

Splunk produces the following result:

```
_time                 count
------------------------
2019-12-04 18:00      2847
2019-12-05 06:00      2724
2019-12-05 18:00      3049
2019-12-06 06:00      2701
2019-12-06 18:00      3205
2019-12-07 06:00      2625
2019-12-07 18:00      3160
2019-12-08 06:00      2418
2019-12-08 18:00      2948
2019-12-09 06:00      2590
2019-12-09 18:00      2880
2019-12-10 06:00      2579
```

2019-12-10 18:00	3117
2019-12-11 06:00	2578
2019-12-11 18:00	111

In the option span=12h, the letter h denotes hour, which is the *timescale identifier*. Table 3-2 lists the *timescale identifiers* you can use.

Table 3-2. Splunk timescale syntax

Timescale	Valid syntax
Second	s | sec | secs | second | seconds
Minutes	m | min | mins | minute | minutes
Hour	h | hr | hrs | hour | hours
Day	d | day | days
Week	w | week | weeks
Month	mon | month | months
Subseconds	us | ms | cs | ds

Using Aggregation Functions

You can use all of the stats aggregation functions with timechart. For example, to plot the 95th percentile value for the *bytes* field from the Splunk tutorial data

```
index=main sourcetype=access_combined_wcookie
| timechart span=1h perc95(bytes) AS "95th precentile request size"
```

Splunk produces the visualization as shown in Figure 3-4.

CHAPTER 3   USING TIME-RELATED OPERATIONS

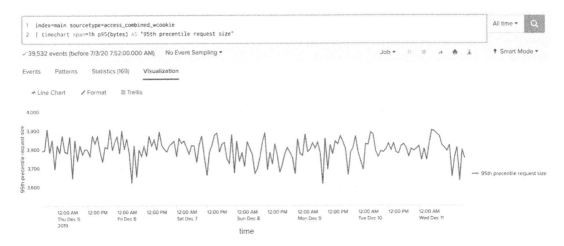

***Figure 3-4.** A line chart created using timechart with p95 aggregation function*

The most common aggregation functions are the following:

- sum
- avg
- count
- distinct_count
- min
- max
- median
- range

In addition, you can use the following functions with `timechart`:

- per_day
- per_hour
- per_minute
- per_second

The preceding *per* functions calculate the *rate* of a metric. For example, if you want to find the total amount of bytes transferred per hour

```
index=main sourcetype=access_combined_wcookie
| timechart per_hour(bytes) AS "Rate of bytes transferred"
```

Splunk produces the following result:

```
_time              Rate of bytes transferred
----------------------------------------
2019-12-04         105155.291667
2019-12-05         509179.375000
2019-12-06         518113.833333
2019-12-07         511911.291667
2019-12-08         462761.250000
2019-12-09         483386.583333
2019-12-10         483731.791667
```

Note that the per_* functions are not responsible for setting the time span. The *per_hour* rate is calculated as *sum()/number of hours in the time span. Either* the time span is specified by you or Splunk chooses the default time span based on the time range of the search.

In addition to the `stats` functions, you can also use `eval` functions for aggregating the values. For example, to plot the average kilobytes over time

```
index=main sourcetype="access_combined_wcookie"
| timechart eval(avg(bytes)/1024) AS Average_Kilo_Bytes
```

The preceding SPL uses `eval` to calculate the average bytes and divide that by 1024 to derive the kilobytes. Note that when you use eval in this fashion, a dynamic field is created by Splunk, so it must always be renamed using the AS clause.

Using Split-by Fields

You can specify *one* split-by field with `timechart`. Note that while `stats` allows unlimited split-by fields, `timechart` allows only one. Also note that the command `chart` allows two split-by fields where the first field used for x axis and second field for y axis.

CHAPTER 3   USING TIME-RELATED OPERATIONS

For example, the Splunk tutorial data has a field named *action* in the events with sourcetype *access_combined_wcookie*. This field denotes the action performed by the user, such as *addtocart, purchase, view,* etc. The following SPL plots the number of actions performed over time, grouped by the action performed:

```
index=main sourcetype=access_combined_wcookie action=*
| timechart count by action
```

Splunk produces the visualization as shown in Figure 3-5.

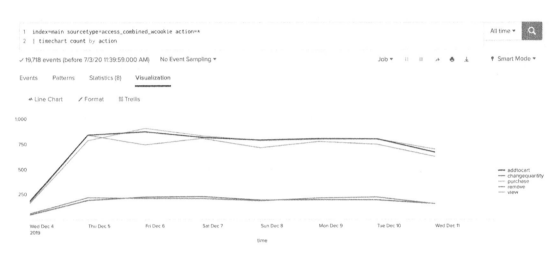

Figure 3-5. *Using timechart with split-by field*

Note that I've used action=* to exclude events with null values for the action field. Instead, you can also use the usenull=f option after the by clause as follows to produce the same result:

```
index=main sourcetype=access_combined_wcookie
| timechart count by action usenull=f
```

Let's say you want the events with NULL action also to be plotted, but you want to use some meaningful label instead of NULL. You can do so with nullstr option as follows:

```
index=main sourcetype=access_combined_wcookies
| timechart count by action nullstr="No action performed"
```

Splunk produces the visualization as shown in Figure 3-6.

CHAPTER 3   USING TIME-RELATED OPERATIONS

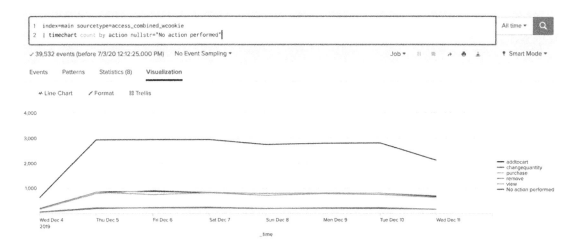

Figure 3-6. *Using timechart with custom nullstr*

In addition, timechart command by default limits the number of values of the split-by field to *ten highest values* (same as chart command). The remaining values are grouped into OTHER. If you don't want the OTHER field to be created, use the option *useother=f*. For example:

```
index=main sourcetype="access_combined_wcookie"
| timechart count by productId usenull=f useother=f
```

Note that the preceding command still displays only the top ten values (default) for the *productId* field, but it won't group the remaining values into OTHER field. Instead if you want all the possible values to be plotted, set the limit option to 0:

```
index=main sourcetype="access_combined_wcookie"
| timechart limit=0 count by productId usenull=f useother=f
```

The preceding SPL will plot one series for each *productId* regardless of how many unique *productIds* are retrieved in the search results.

Basic Examples

Let us take a look at a few basic applications of timechart. In any web application, monitoring the HTTP errors is a vital task. Using timechart, you can plot the hourly count of HTTP errors and visualize in a stacked column chart. Using the Splunk tutorial data

```
index=main sourcetype="access_combined_wcookie" status != 200
| timechart count by status useother=f
```

91

CHAPTER 3 USING TIME-RELATED OPERATIONS

In the resulting visualization as shown in Figure 3-7, click the current visualization selected and choose *column chart.*

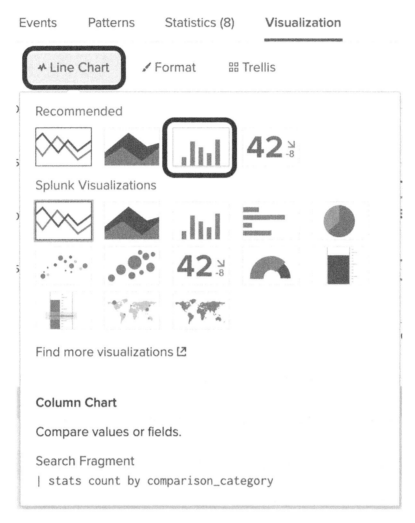

Figure 3-7. *Selecting column chart visualization*

In the resulting column chart, click *Format,* and in the *General* tab, click stacked mode as shown in Figure 3-8.

CHAPTER 3   USING TIME-RELATED OPERATIONS

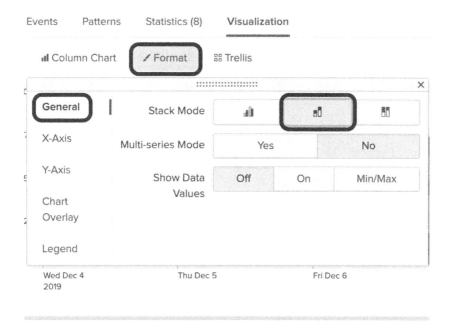

Figure 3-8. Selecting column stack mode

The resulting stacked column chart provides an insightful representation of the distribution of the HTTP status codes. See Figure 3-9.

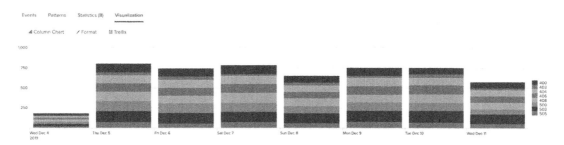

Figure 3-9. A column chart in stacked mode generated by timechart

Note   In practice, any 200 family or 300 family HTTP status codes will be acceptable as success. Depending on the HTTP status codes found in your log files, you may have to add additional filtering to produce accurate reporting.

Chapter 3 Using Time-Related Operations

Another useful application of timechart is to plot two or more related metrics in one graph to gain additional insight. For example, using the same Splunk tutorial data, you can plot successful purchases and failed purchases as follows:

```
index=main sourcetype="access_combined_wcookie"
| timechart span=1h
    count(eval(action = "purchase")) AS "Total purchases"
    count(eval(status != 200 AND action = "purchase")) AS "Failed purchases",
    count(eval(status = 200 AND action = "purchase")) AS "Successful purchases"
```

Splunk produces the following visualization (Figure 3-10) viewed as a line chart.

Note that I've made use of eval expressions to determine successful purchases and failed purchases. Using eval expressions provides a lot of flexibility due to the numerous functions you can use with it.

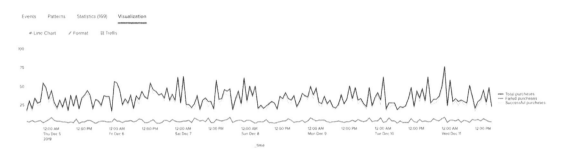

Figure 3-10. Using timechart to plot relevent series in the same graph

Additional Useful Tips

In this section, I'll point out some useful tips related to time that are usually overlooked. These are not necessarily SPL commands. These are shortcuts you can use in Splunk Web interface, or automatically extracted fields that you may not be aware of. Let's take a look.

CHAPTER 3   USING TIME-RELATED OPERATIONS

Retrieving Events in Time Proximity

At times, you will want to retrieve logs that were created just before a certain event, such as a failed transaction. You may also want to retrieve logs that were created after that particular event. There are two ways to do this easily. The first option is to simply retrieve the raw logs in the Splunk Web interface. To do this, first retrieve the event you are interested in. In the following example, I'm searching for a particular user session where the user removed some item from the cart:

index=main sourcetype="access_combined_wcookie" JSESSIONID=SD10SL8FF5ADFF31078 action=remove

In the results, once you identify the event you are interested in, simply expand the event by clicking the > character at the left, click *Event Actions,* and choose *Show Source.* See Figure 3-11.

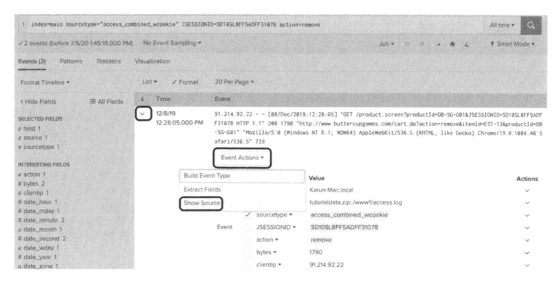

Figure 3-11. *Showing raw data from the search results*

Splunk opens the raw logs in a separate browser window with the event that you click on highlighted. This immediately reveals the log entries above and below the event you are interested in. See Figure 3-11.

CHAPTER 3   USING TIME-RELATED OPERATIONS

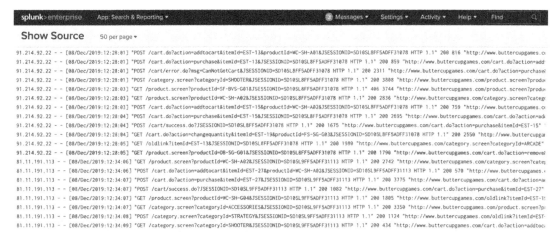

Figure 3-12. Raw data displayed in Splunk Web

The second option is to use *Nearby Events* menu from the event. In order to access this menu, simply click the timestamp of the event you are interested in and select the time range. See Figure 3-13.

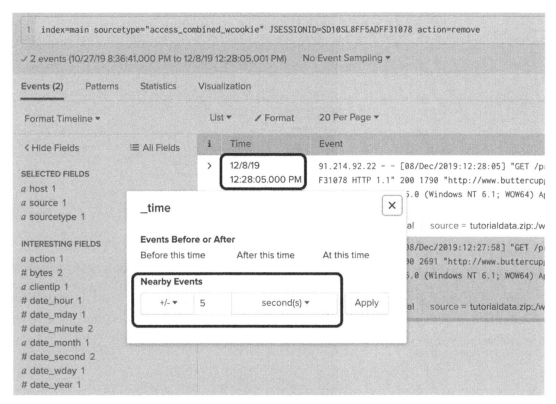

Figure 3-13. Selecting nearby events in Splunk Web

CHAPTER 3　USING TIME-RELATED OPERATIONS

Once you select the desired time range and click apply, Splunk will rerun the search with a new time range. You may want to remove any filtering you had in your SPL and rerun if you want to retrieve all logs during the time range selected.

Using the date_time Fields

Splunk automatically extracts the following very useful time-related fields for every event:

- date_second
- date_minute
- date_hour
- date_mday
- date_month
- date_year
- date_wday
- date_zone

These fields should automatically appear on the left side under *INTERESTING FIELDS* when your search mode is *smart mode* or *verbose mode* (in *fast mode*, automatic field discovery is disabled). See Figure 3-14.

```
INTERESTING FIELDS
a  action  5
#  bytes  100+
a  categoryId  8
a  clientip  100+
#  date_hour  24
#  date_mday  5
#  date_minute  60
a  date_month  1
#  date_second  60
a  date_wday  5
#  date_year  1
a  date_zone  1
#  duration  100+
a  file  14
```

Figure 3-14. *Automatically extracted date_time fields*

CHAPTER 3 USING TIME-RELATED OPERATIONS

There are many practical applications of these automatically extracted fields. For example, if you want to plot a metric only during business hours (9 AM to 5 PM), you can easily do so by utilizing the *data_hour* field. The *date_hour* field stores the hour in 24-hour clock. Using the Splunk tutorial zip

```
index=main sourcetype="access_combined_wcookie" date_hour >= 9 AND date_hour < 17
| stats count AS "Invocations during business hours" by action
```

The preceding SPL calculates the number of invocations grouped by the type of action, ignoring the off-business hours.

You can also calculate the hourly average of a metric. For example, to find the total number of hits to the website per hour in a 24-hour clock

```
index=main sourcetype="access_combined_wcookie"
| stats count by date_hour
| sort date_hour
```

Splunk produces the following visualization in column chart. See Figure 3-15.

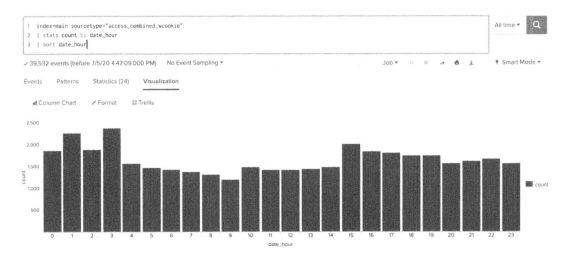

Figure 3-15. *Using date_hour field to plot hourly*

As another example, you can calculate the weekend traffic by using the date_wday field:

```
index=main sourcetype="access_combined_wcookie" date_wday IN
("saturday","sunday")
| stats count AS "Invocations during weekends" by action
```

The preceding SPL retrieves events that occurred during weekends (Saturday or Sunday) and pipes the results into `stats` command to produce a report of the number of invocations grouped by *action*.

Caution Your events should have valid timestamps in order for the *date_time* extractions to work.

Using Time Modifiers

The *time picker* in the search interface allows you to select the time range of your search in a variety of ways. Using the *time-picker* menu, there are several options to choose from: preset, real time, relative time, date and time range, and so on. But Splunk provides another powerful way to specify the time range using SPL itself – *time modifiers*. Using time modifiers allows you to customize the time range as part of the SPL query itself. This also comes in handy when you execute your SPLs using REST-API where you don't have access to time-picker interface. When you use time modifiers in an SPL, they override any time range set in the time range picker.

You can use the `earliest` and `latest` keywords in SPL to specify the time modifiers. The keyword `earliest` specifies the earliest _time for the time range of your search, while the keyword `latest` specifies the latest _time for the time range. Consider the following example:

```
index=_internal sourcetype=splunkd earliest=-24h
```

In the preceding SPL, time range is set to the previous 24 hours using the key-value pair `earliest=-24h`. You can specify the time using a number followed by a unit. To make the time relative to now, you can add + or – in front of the number. This custom time range will override any time range specified using the time picker. Table 3-3 lists the units you can use.

Table 3-3. Splunk time units and their abbreviations

Time unit	Unit abbreviation
Millisecond	ms
Second	s, sec, secs, second, seconds
Minute	m, min, minute, minutes
Hour	h, hr, hrs, hour, hours
Day	d, day, days
Week	w, week, weeks
Month	mon, month, months
Quarter	q, qtr, qtrs, quarter, quarters
Year	y, yr, yrs, year, years

The sign in the number specifies whether to go backward or forward. For example, 2h indicates two hours ago, and +1h indicates one hour ahead. You can omit the number for single time amount. So, -1d and -d both specify one day ago. You can also specify the exact time such as `earliest="4/12/2020:10:05:00"`.

Specifying a Snap-to Time Unit

A handy feature of time modifiers is the ability to snap to the time unit to round the time to the nearest value. You use the character @ to separate the time unit and the snap-to time unit. Here are some examples.

To set the time range for the past hour

`earliest=-1h@h latest=@h`

If the current time is 9:25 PM, the preceding time modifier will result in the time range from 8:00 PM to 9:00 PM. Note that if you omit the number 1, Splunk will assume 1.

To set the time range for the past two days

`earliest=-2d@d latest=@d`

If the current time is 9:25 PM on July 5, the preceding time modifier will result in the time range from 12:00 AM on July 3 through 12:00 AM on July 5.

CHAPTER 3　USING TIME-RELATED OPERATIONS

To the time range from the start of the week until now

`earliest=-1w@w latest=now`

If the current time is 9:25 PM on Friday, the preceding time modifier will result in the time range from 12:00 AM on previous Monday through now. The time modifier now is a special time that takes the value of the time when the search is executed.

Note When Splunk snaps to the nearest time, it always snaps backward. It never rounds to the latest time after the specified time. For example, if the current time is 9:45 PM, the time modifier @h results in 9:00 PM. Similarly, the time modifier @d results in 12:00 AM of the current day.

You can also specify an offset for the snap-to time unit. For example, to set the time range from 2:00 AM yesterday to 2:00 AM today, you can use the following time modifier:

`earliest=-1d@d+2h latest=@d+2h`

Similarly, to set the time range from 14th 12:00 AM of the previous month to 14th of the current month 12:00 AM, you can use the following time modifier:

`earliest=-1mon@mon+13d latest=@mon+13d`

One shortcut I've found useful when it comes to specifying the correct time modifier is to use the time picker's *advanced* time range to verify the expression I'm trying to write. Figure 3-16 shows how using time picker can help to verify the resulting time frame based on the expression you type in the `earliest` and `latest` fields.

CHAPTER 3   USING TIME-RELATED OPERATIONS

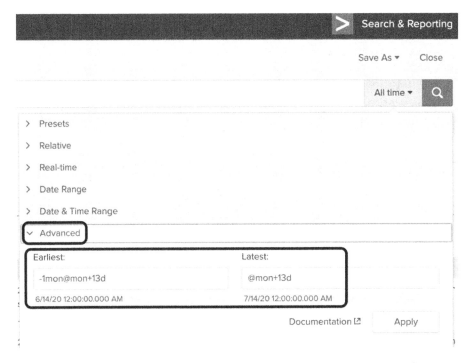

Figure 3-16. Utilizing the time picker to validate custom time modifiers

Advanced Examples

Now that you've learned about `timechart` command and how you can use time modifiers, let us take a look at a few advanced examples where we can make use of time-related features in Splunk to bring about useful insights from your data.

Comparing Different Time Periods

In IT operations, one of the widely used troubleshooting tactics is to compare a particular metric during different time frames. For example, if the response time of an application is very high on a Monday, it makes sense to compare the response time from the same time last Monday, and perhaps the Monday before as well. Similarly, you may want to compare a metric between first day of current month and first day of previous month and so on. Splunk provides a SPL command named `timewrap` to easily achieve this kind of day-over-day or month-over-month comparisons.

CHAPTER 3 USING TIME-RELATED OPERATIONS

The `timewrap` command wraps the time, so to speak, based on the time range you provide, in order to plot different time frames as *individual series* in a timechart. Consider the following example:

```
index=_internal sourcetype=splunkd log_level=WARN
| timechart count span=1h
```

The preceding SPL retrieves events with *log_level* equals to *WARN* from *_internal* index and plots the data in a chart. See Figure 3-16.

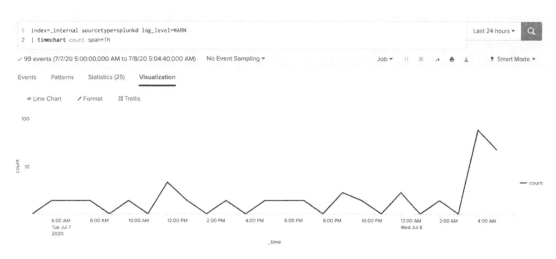

Figure 3-17. *Using timechart to plot the number of events over time*

Now, let us say you want to compare this with the previous three days. You could easily achieve this with `timewrap` command by piping the results of the preceding command to `timewrap 1d` as shown in the following:

```
index=_internal sourcetype=splunkd log_level=WARN
| timechart count span=1h
| timewrap 1d
```

You must set the time range of the search to the number of days you want to go back. The 1d in the `timewrap` command is a required argument, and it denotes the wrapping span. Splunk produces the graph as shown in Figure 3-18. As you can see, `timewrap` command creates different series named `latest_day`, `1day_before`, `2days_before`, and so on.

103

CHAPTER 3   USING TIME-RELATED OPERATIONS

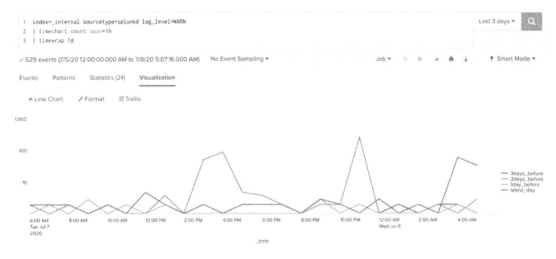

Figure 3-18. Using timewrap to multiple time series

Note   The `timewrap` command should always be preceded by the `timechart` command.

The wrapping span can be specified using an integer followed by the timescale option such d for days, m for months, w for weeks, and so on. Note that the `timewrap` command uses m for months whereas `timechart` uses m to refer minutes.

You can customize the label generated for the different time series by using the `series` option. The *series* option takes one of the following three values:

- relative (this is the default option)
- exact
- short

For example, consider the following SPL:

index=_internal sourcetype=splunkd log_level=WARN
| timechart count span=1h
| timewrap 1d **series=exact**

The time series generated by the preceding SPL would be based on the exact date of time. See Figure 3-19.

CHAPTER 3   USING TIME-RELATED OPERATIONS

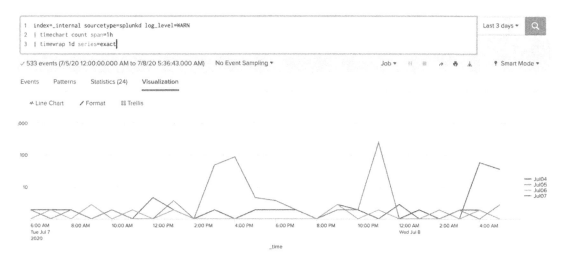

Figure 3-19.   Using the series option with timewrap

The series option short results in time series generated as s0,s1,s2, and so on. When you use exact series, you can also generate custom names with the time_format option.

Another application of timewrap is to compare today, yesterday, and average for the past seven days. Consider the following SPL:

```
index=_internal sourcetype=splunkd log_level=WARN
| timechart count span=1h
| timewrap 1d
| addtotals
| eval weeklyAvg = Total/7
| table _time,latest_day,1day_before,weeklyAvg
```

The preceding SPL uses timewrap to first generate day-over-day statistics. It then adds the daily totals to find the weekly average using addtotals and eval. Finally, it plots just the current day, yesterday, and weekly average using the table command.

Tip   The command addtotals computes the arithmetic sum of all numerical fields and stores the results in a new filed named Total.

105

CHAPTER 3 USING TIME-RELATED OPERATIONS

Comparing the Current Day with Average of the Past 30 Days

Occasionally, you may want to compare the current-day metrics to the average of the past few days, typically the past 30 days. While calculating averages, an important point to note is that you don't want to simply average the entire 30 days without taking *hour of the day* into consideration. For example, you want to compare the metrics between 9:00 AM and 10:00 AM of the current day with the average of the *same hour* over the past 30 days. At first, this may be tricky to understand, but the key is to split the metrics by hour of the day. You would also need to create two series, one for current day and one for the average. The algorithm for SPL will be as follows:

- Retrieve events for the past 30 days.
- Mark today's events as today and previous 30 day's events as monthly by creating new field named *series*.
- Use timechart to calculate the count per hour (span=1h) by the series.
- Create a new field *hour* by using strftime function on the field *_time*.
- Use stats to calculate the average count by hour on the two series.

For this example, let us calculate the number of WARN messages in *_internal* index. Let us walk through the SPL:

```
index=_internal log_level=WARN earliest=-30d@d latest=now()
```

The preceding SPL simply retrieves all the WARN events from _internal index for the past 30 days. The now() function retrieves the current time:

```
index=_internal log_level=WARN earliest=-30d@d latest=now()
| eval series = if(_time > relative_time(now(),"@d"), "today","monthly")
```

Now, we create a new field named *series* and use the eval's if function to assign the value *today* if *_time* of an event is greater than 12:00 AM of current day. The function relative_time(now(),"@d") applies the time modifier @d to now() to derive 12:00 AM of the current day. If *_time* of an event is greater than 12:00 AM of current day, the value *today* is assigned. Otherwise, the value *monthly* is assigned. If you simply run the preceding SPL and examine the fields extracted, you would notice the field *series* with two values. See Figure 3-20.

CHAPTER 3   USING TIME-RELATED OPERATIONS

Figure 3-20. Creating multiple series based on time

```
index=_internal log_level=WARN earliest=-30d@d latest=now()
| eval series = if(_time > relative_time(now(),"@d"), "today","monthly")
| timechart count span=1h by series
```

Here, we use the `timechart` command to plot the count with the span of one hour and split the data by the *series*. An excerpt of the output of the preceding SPL in the Statistics tab looks like the following:

```
_time                monthly    today
-----------------------------------
2020-07-09 01:00     0          24
2020-07-09 00:00     0          274
2020-07-08 23:00     319        0
2020-07-08 22:00     2          0
2020-07-08 21:00     1          0
```

107

CHAPTER 3   USING TIME-RELATED OPERATIONS

Let us continue building our SPL:

```
index=_internal log_level=WARN earliest=-30d@d latest=now()
| eval series = if(_time > relative_time(now(),"@d"), "today","monthly")
| timechart count span=1h by series
| eval Hour = strftime(_time,"%H")
```

Here, we extract the hour from *_time* into a new field named *Hour*. An excerpt of the output of the preceding SPL in the Statistics tab looks like the following:

```
_time                Hour   monthly    today
-----------------------------------------
2020-07-09 01:00     01     0          24
2020-07-09 00:00     00     0          274
2020-07-08 23:00     23     319        0
2020-07-08 22:00     22     2          0
2020-07-08 21:00     21     1          0
```

```
index=_internal log_level=WARN earliest=-30d@d latest=now()
| eval series = if(_time > relative_time(now(),"@d"), "today","monthly")
| timechart count span=1h by series
| eval Hour = strftime(_time,"%H")
| stats avg(monthly) AS MonthlyAverage, sum(today) AS Today by Hour
```

Finally, we use the `stats` command to calculate average by the *Hour* for *monthly* and *today*. Note that we use `sum(today)` as there are only one set of values for a given hour and hence taking average does not make sense. An excerpt of the output of the preceding SPL in the Statistics tab looks like the following:

```
Hour    MonthlyAverage          Today
-----------------------------------
00      1.1935483870967742      274
01      0.3870967741935484      24
02      1.6774193548387097      0
03      0.6774193548387096      1
04      3.2903225806451615      0
```

The line chart generated in the Visualization tab is shown in Figure 3-21.

108

CHAPTER 3 USING TIME-RELATED OPERATIONS

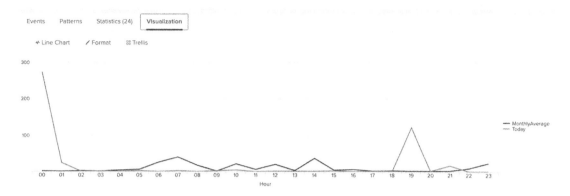

Figure 3-21. *Comparing today with 30 days average*

As evident from the preceding chart, the hours 00 and 19 are clearly problematic in the *today* series, as the 30-day average during those hours were much smaller. This is the kind of insight you can generate from your operational logs by using SPL.

Using Time Arithmetic

Using SPL, you can perform arithmetic operations on time by adding or subtracting seconds. Note that this doesn't actually change the time of the event but enables us to use the time tailor our searches. This can come in handy if you want to mimic certain time ranges as something else, for example, to plot yesterday's metrics alongside day-before-yesterday's metrics. Let us consider an example where you want to plot the *sales by the hour* of your business. Assume that *sales* is a field in your logs. You want to compare the yesterday's numbers with day-before-yesterday. The algorithm for SPL will be as follows:

- Retrieve all events for the past two days, using earliest=-2d@d and latest=@d.

- Mark the yesterday's events as *Yesterday* and day-before-yesterday's events as *Day-before-yesterday* by creating a new field named *marker* using `eval` and `if`.

- Update _time of Day-before-yesterday's events to look like yesterday's events. You do this by adding 24*60*60 seconds to _time.

- Use `timechart` to calculate the sum of sales with the span of one hour and split by the marker.

109

The SPL excerpt will look like the following:

```
...  earliest=-2d@d latest=@d
| eval marker = if(_time > relative_time(now(),"-1d@d"),"Yesterday","Day-before-yesterday")
| eval _time = if(marker == "Day-before-yesterday", _time + 24*60*60, _time)
| timechart sum(sales) span=1h by marker
```

You can use `eval` and `if` to determine if *_time* falls under yesterday or day-before-yesterday. You utilize the `relative_time` function of eval which takes a Unix time as the first argument and a time identifier as the second argument. It applies the time modifier to the Unix time and returns the resulting time. In our example, the expression `relative_time(now(), "-1d@d")` will result in 12:00 AM of the previous day. We then add 1 day (24 * 60 * 60) in seconds if the time of the event is day-before-yesterday. Finally, we use `timechart` to plot the sum of sales with the span of 1h and split by the marker. The resulting `timechart` will have two series, one for yesterday and one for day-before-yesterday. Note that we could have easily achieved this by using `timewrap` command we discussed earlier. But this example reveals the power you have to manipulate time in SPL.

Key Takeaways

In this chapter, you learned the various ways you can handle time in Splunk. Time is the most important component of Splunk's data architecture. Every event that is indexed in Splunk has an associated timestamp. Visualizing data plotted over time is an important step in troubleshooting an impaired system or application. Splunk provides numerous SPL commands and `eval` functions to effectively make use of time. Here are the key takeaways from this chapter:

1. In order to make the best use of Splunk, the raw data from your log files and other data sources must contain a valid timestamp including the time zone.

2. You can explicitly set your time zone in Splunk Web's user preferences.

3. Internally, Splunk always stores timestamps in Unix time, which is the number of seconds since midnight on January 01, 1970.

4. You can use the SPL command `timechart` to plot a metric over time.

5. `Timechart` allows only one split-by field. Each unique value of the split-by field becomes a series in the resulting chart.

6. You can use the `span` option to specify the size of discrete bins in the timechart.

7. You can use the automatically extracted time fields such as *date_mday*, *date_hour,* and so on.

8. By utilizing `eval` and `relative_time`, you can create multiple time series to be plotted on the same chart for comparing.

9. You can perform arithmetic operations with time by adding or subtracting the number of seconds.

10. You can utilize the handy `timewrap` command to compare multiple time frames.

That concludes the discussion on time-related operations in Splunk. By utilizing the `timechart` command and its various options, you can gain lots of insights from your machine data. In addition, you can make use of the `timewrap` command to compare various time frames. But learning how to use `stats`, `eval`, and `timechart`, you are in a great position to make the best of the Splunk platform. As you start using Splunk for many of your operational data intelligence needs, you will invariably find the need to group and correlate data across multiple data sources. SPL provides a rich set of commands to group and correlate the events. In the next chapter, we'll discuss grouping and correlating commands of SPL.

CHAPTER 4

Grouping and Correlating

When you collect your log data from multiple data sources such as network devices, servers, and applications, the need for correlating and grouping those logs may raise. For example, your application server log might store the transaction ID in the application server's log files. If your application utilizes an external service, which is not uncommon, the transaction ID might appear on its log files. If you want to know the complete end-to-end activities of a particular transaction ID, you need to correlate your application server's logs with the external service's log files. Splunk's SPL provides a rich set of commands to group data from multiple sources.

In this chapter, let us take a look at the SPL commands that are specifically designed to group and correlate data from various sources. We will review the `transaction` command which is used to group related events using the constraints you specify. We will then learn the `join` command which acts similar to the SQL join. We'll also study the `append` family of commands that lets you add results from a subsearch to the main search. The term *grouping* and *correlating* can be interchangeably used throughout this chapter.

> **Note** We are not discussing the correlation search as defined in Splunk Enterprise Security add-on. Splunk Enterprise Security is a premium Splunk app that is purpose-built to address security use cases.

Transactions

The command `transaction` is used to group conceptually related events, even if they are from multiple data sources. You provide one or more constraints to define the criteria for grouping. The events are grouped into transactions. A transaction contains the following items:

- Raw events based on the criteria provided by you
- A union of fields from all the events in the transaction
- A field named `duration` that denotes the time elapsed between the first and last events of the transaction
- A field named `eventcount` that denotes the number of events in the transaction
- A field named `closed_txn` that denotes whether the transaction is a closed one (complete) or evicted (incomplete)

The timestamp of the earliest event in the transaction becomes the timestamp of the transaction. The basic syntax of the `transaction` command is as follows:

```
| transaction <constraints>
```

You simply pipe the results of a search into `transaction` command with the constraints you need for the grouping. Let us dive into the `transaction` command and its usage.

Using Field Values to Define Transactions

The simplest form for creating a transaction is to use a field as a constraint. Consider the following example which uses the Splunk tutorial data:

```
index=main sourcetype="access_combined_wcookie"
| transaction JSESSIONID
```

The preceding SPL query retrieves events from *main* index with sourcetype *access_combined_wcookie* and pipes the results into the `transaction` command. The only parameter to the `transaction` command is a field named *JSESSIONID*. This field must be present in the events in order for the event to be considered to be part of a

CHAPTER 4 GROUPING AND CORRELATING

transaction. The SPL query groups the raw events that have the same value for the field JSESSIONID into transactions. JSESSIONID is a unique ID (cookie) given to a user session in the web application (it is a Java Servlet terminology). By using JSESSIONID, one can identify a particular user session in the web access logs. The result will look like Figure 4-1.

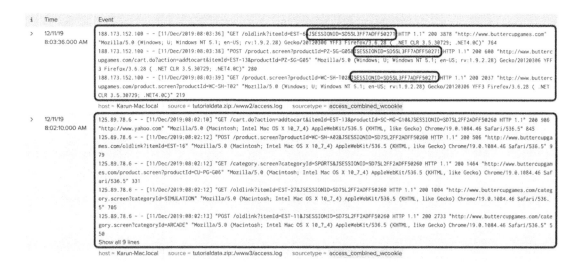

Figure 4-1. Output of transaction command

As you can see, events with the same JSESSIONID are grouped as a transaction. The timestamp of the transaction is the timestamp of the earliest event. Note that the events within the transactions are ordered *chronologically*, which puts the earliest event first. In addition to grouping the events as transactions, the transaction command also creates some fields. These fields can be seen in the *fields sidebar*. See Figure 4-2.

CHAPTER 4 GROUPING AND CORRELATING

Figure 4-2. Fields generated by transaction command

The field *duration* is of particular interest. It calculates the time elapsed between the first and last events within the transaction in seconds. For example, if you want to retrieve transactions that lasted for more than 15 seconds, you can use the following SPL:

```
index=main sourcetype="access_combined_wcookie"
| transaction JSESSIONID
| search duration >= 15
| table _time, JSESSIONID, duration
```

The SPL query uses the autogenerated field *duration* to check if the duration is more than 15 seconds and then uses the `table` command to print just the relevant fields. The output will look like this:

```
_time                   JSESSIONID              duration
-----------------------------------------------------------
2019-12-09 07:16:47     SD1SL7FF2ADFF36483      15
2019-12-06 21:56:29     SD2SL4FF3ADFF19958      15
2019-12-05 08:26:23     SD9SL1FF6ADFF9130       15
```

As you can see, there are 3 transactions that lasted for 15 seconds or more. In addition, you can also use the max function of stats to find the maximum duration among all transactions as follows:

index=main sourcetype="access_combined_wcookie"
| transaction JSESSIONID
| **stats max(duration)**

The maximum duration happens to be 15 seconds, as seen from the following output:

```
max(duration)
-------------
15
```

You can use more than one field to define the transaction by a unique combination of fields. This may be desirable in some conditions. For example, to group the events based on *JSESSIONID* and *uri_path*

index=main sourcetype="access_combined_wcookie"
| **transaction JSESSIONID, uri_path**

The preceding SPL will group events based on the unique combination of *JSESSIONID* and *uri_path*.

Using Strings to Define Transactions

You can also group events using string that are present in the *start* or *end* of a transaction. You can use the option *startswith* to specify a string that must be present in the first event of the transaction. Similarly, you can use *endwith* to specify a string that must be present in the last event of the transaction. You can use either *startswith* or

CHAPTER 4   GROUPING AND CORRELATING

endswith, or *both*. Further, instead of literal strings, you can use any eval expression for these options. For example, in order to group events that occurred during a *purchase* by a given user, you could use the following SPL:

```
index=main sourcetype="access_combined_wcookie"
| transaction JSESSIONID, clientip startswith="action=view"
endswith="action=purchase"
```

During the execution of the SPL query, first all events from *main* index with sourcetype *access_combined_wcookie* are retrieved. The results are piped into the transaction command that groups events using the following constraints:

- The events must share the same combination of JESSIONID and client_ip. This ensures that the transaction is for a unique user session.

- The start of the transaction must be an event that has the string "action=view".

- The end of the transaction must be an event that has the string "action=purchase".

Figure 4-3 shows one transaction that has been generated using this SPL.

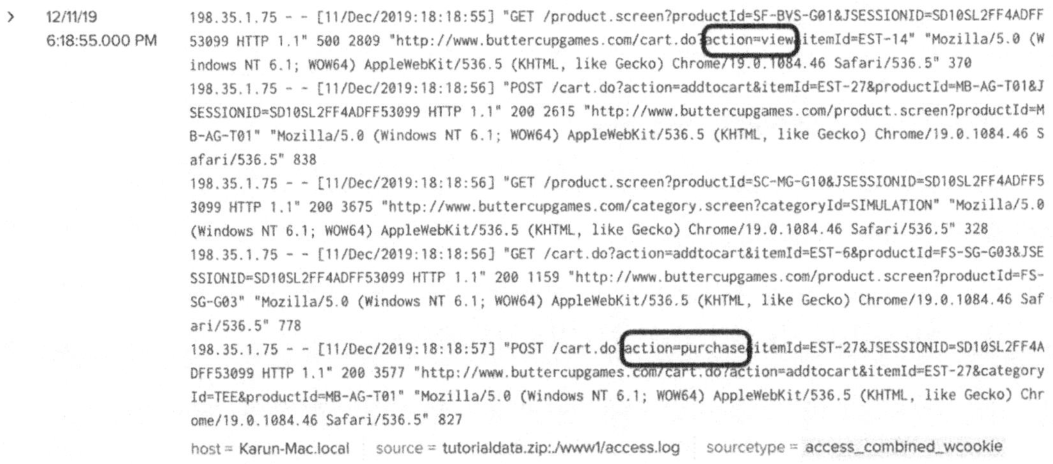

Figure 4-3. Output of transaction command

As you can see, the first event has the string *"action=view"*, and the last event has the string *"action=purchase"*. This transaction produces an easily understandable view of the events that happened during the purchase. Now that you have the transaction defined, you can use this for other tasks. For example, you can calculate the average shopping time, that is, the time elapsed between a customer viewing an item and purchasing it eventually. The following SPL will do the job:

```
index=main sourcetype="access_combined_wcookie"
| transaction JSESSIONID, clientip startswith="action=view"
endswith="action=purchase"
```
| stats avg(duration) AS "Average shopping time"

Splunk produces the following output:

```
Average shopping time
---------------------
2.932242990654206
```

Let us consider another example. In Splunk *_internal* logs, you will find *splunkd* startup logs. *Splunkd* is the primary Splunk process that handles indexing and searching. Let's say you want to find out how long does it take for *splunkd* to start up during restarts. You can use an SPL like in the following:

```
index=_internal sourcetype=splunkd
```
| transaction startswith="My server name is" endswith="UiHttpListener - Limiting UI HTTP"

First, the SPL query retrieves data from *_internal* index with sourcetype *splunkd*. Then, it pipes the results to the `transaction` command that defines two constraints as follows:

- The start of the transaction must be an event that has the string "My server name is".

- The end of the transaction must be an event that has the string "UiHttpListener - Limiting UI HTTP".

Now, you can produce a useful report like the following:

```
index=_internal sourcetype=splunkd
```

CHAPTER 4 GROUPING AND CORRELATING

```
| transaction startswith="My server name is" endswith="UiHttpListener -
Limiting UI HTTP"
| stats count AS "Number of splunkd restarts", avg(duration) AS "Average
startup time in seconds"
```

The SPL query calculates the number of restarts and the average startup time for the time range you specify in the time picker. The result is shown in the following:

```
Total splunkd restarts    Average startup time in seconds
-----------------------------------------------------------
9                         0.6248888888888889
```

Caution The log messages found in _internal index may be changed during Splunk version upgrades. So, use this SPL example with caution.

Using Additional Constraints

In addition to field values, *startswith* and *endswith*, you can also use a few other useful constraints with `transaction` command. These constraints provide even more ways for you to define the grouping of your events. You can use these constraints along with other constraints in the same `transaction` command. The constraints are

- maxspan
- maxpause
- maxevents

Let us look at each of them in detail.

maxspan

The constraint *maxspan* defines the maximum length of time in seconds, minutes, hours, or days that the events within a transaction can span. The events that exceed this limit are not grouped into the current transaction. The time unit s, m, h, and d are used to specify seconds, minutes, hours, and days, respectively. For example:

```
index=main sourcetype="access_combined_wcookie"
| transaction JSESSIONID, clientip maxspan=5m
```

CHAPTER 4 GROUPING AND CORRELATING

The preceding query first retrieves all events from main index with sourcetype *access_combined_wookie*. It then pipes the results to the `transaction` command which groups the events which share the same *JSESSIONID* and *clientip*, as long as the total duration (span) of the transaction does not exceed five minutes. The total duration is the elapsed time between the first and the last events in the transaction. Events that exceed five minutes are treated for the next transaction.

maxpause

The constraint *maxpause* defines the maximum amount of time that can exist *between* the events within a transaction. If the time elapsed between two events exceeds this limit, they are not grouped into a transaction. Similar to *maxspan*, you specify the time in terms of seconds, minutes, hours, or days. This constraint can come in handy when you don't want to group events with big gaps between them. For example, consider the following SPL:

```
index=main sourcetype="access_combined_wcookie"
| transaction JSESSIONID, clientip maxspan=5m maxpause=30s
```

The `transaction` command not only imposes the fields *JSESSIONID* and *clientip* fields to be common among the events but also ensures the maximum duration (span) of the transaction is less than 5 minutes and no 2 consecutive events occurred more than 30 seconds apart.

maxevents

The constraint *maxevents* defines the maximum number of events a transaction can be made of. This constraint can be handy if you want to put a cap on the number of events:

```
index=main sourcetype="access_combined_wcookie" action="purchase"
| transaction JSESSIONID, clientip maxpause=30s maxevents=5
```

The preceding query first retrieves all the purchase events from main index with sourcetype *access_combined_wookie*. It then uses `transaction` command to group the events where they share common *JSESSIONID* and *clientip*, enforcing the maximum time elapsed between the events to 30 seconds and the total number of events no more than 5.

CHAPTER 4 GROUPING AND CORRELATING

What Happens to the Fields in a Transaction?

As transactions are usually made up of multiple events, what happens to the fields in those events? For example, consider our earlier example:

```
index=main sourcetype="access_combined_wcookie" action="purchase"
| transaction JSESSIONID, clientip
```

The `transaction` command in the SPL groups purchase events that share common *JSESSIONID* and *clientip*. As you can see from Figure 4-4, which shows one of the transactions, the events within a transaction can have various values for a field.

```
198.35.1.75 - - [11/Dec/2019:18:18:57] "POST /cart.do?action=purchase&itemId=EST-27&JSESSIONID=SD10SL2FF4A
DFF53099 HTTP 1.1" 200 3577 "http://www.buttercupgames.com/cart.do?action=addtocart&itemId=EST-27&category
Id=TEE&productId=MB-AG-T01" "Mozilla/5.0 (Windows NT 6.1; WOW64) AppleWebKit/536.5 (KHTML, like Gecko) Chr
ome/19.0.1084.46 Safari/536.5" 827
198.35.1.75 - - [11/Dec/2019:18:18:57] "POST /cart/success.do?JSESSIONID=SD10SL2FF4ADFF53099 HTTP 1.1" 200
613 "http://www.buttercupgames.com/cart.do?action=purchase&itemId=EST-27" "Mozilla/5.0 (Windows NT 6.1; WO
W64) AppleWebKit/536.5 (KHTML, like Gecko) Chrome/19.0.1084.46 Safari/536.5" 328
198.35.1.75 - - [11/Dec/2019:18:18:58] "POST /cart.do?action=purchase&itemId=EST-16&JSESSIONID=SD10SL2FF4A
DFF53099 HTTP 1.1" 200 821 "http://www.buttercupgames.com/cart.do?action=addtocart&itemId=EST-16&categoryI
d=SIMULATION&productId=SC-MG-G10" "Mozilla/5.0 (Windows NT 6.1; WOW64) AppleWebKit/536.5 (KHTML, like Geck
o) Chrome/19.0.1084.46 Safari/536.5" 178
198.35.1.75 - - [11/Dec/2019:18:18:59] "POST /cart/success.do?JSESSIONID=SD10SL2FF4ADFF53099 HTTP 1.1" 200
2568 "http://www.buttercupgames.com/cart.do?action=purchase&itemId=EST-16" "Mozilla/5.0 (Windows NT 6.1; W
OW64) AppleWebKit/536.5 (KHTML, like Gecko) Chrome/19.0.1084.46 Safari/536.5" 386
```

Figure 4-4. Events in a transaction can have multiple values for a field

This transaction has events that have various values for the field *itemId*. There are two events with the value of EST-27 and two events with the value of EST-16. The `transaction` command simply performs a union of all the field values and stores them in the *itemId* field of the transaction. Let's see the value of *itemId* in this particular transaction. I've added *JSESSIONID* to the search criteria to focus on this particular transaction:

```
index=main sourcetype="access_combined_wcookie" action="purchase" JSESSIONI
D=SD10SL2FF4ADFF53099
| transaction JSESSIONID, clientip
| table itemId
```

The query uses `table` command to print just the *itemId* field. Splunk produces the following output:

```
itemId
-------
EST-16
EST-27
```

As you can see, the duplicate values are removed, and the values do not seem to be in order (in the actual transaction, the item EST-27 came first).

For most practical purposes, this behavior is acceptable. However, if you need to retrieve all the values of a field within a transaction, you can add mvlist=true to the transaction command. Splunk will create a *multivalued* field named *itemId* and keep all the values of the field and also *preserve the order*. Let's rewrite the SPL as follows:

```
index=main sourcetype="access_combined_wcookie" action="purchase"
JSESSIONID=SD10SL2FF4ADFF53099
| transaction JSESSIONID, clientip mvlist=true
| table itemId
```

This query adds mvlist=true to the `transaction` command. This results in *itemId* being created as a multivalued field. The result looks like the following:

```
itemId
-------
EST-27
EST-27
EST-16
EST-16
```

As you can see, the order of the values is maintained, and the values are not deduped.

CHAPTER 4　GROUPING AND CORRELATING

Finding Incomplete Transactions

As stated earlier, using the `transaction` command, you can group conceptually related events into a single transaction based on the constraints you provide. By default, the events that don't qualify to be in a transaction are discarded. Sometimes, you may want to find exactly that: the events that *don't* belong to a transaction. One practical application of this is to find unfinished tractions. For example, consider the following events:

```
[11/Dec/2019:18:18:58] TSM backup_id=12 starting ...
[11/Dec/2019:18:32:25] TSM backup_id=12 %complete=50
[11/Dec/2019:18:45:49] TSM backup_id=12 %complete=75
[11/Dec/2019:19:03:11] TSM backup_id=12 completed.
```

The events show the progress of a backup job. The backup job starts with the string *"starting"* and ends with the string *"completed."* There is also a field named *backup_id* that indicates the backup ID. In order to group the preceding events as transaction, the following SPL can be employed:

```
index="main" sourcetype="backup:log"
| transaction backup_id startswith="starting" endswith="completed"
```

The `transaction` command in the SPL query groups the events based on the value of *backup_id* field, marking the first event of the transaction with the string *"starting"* and the last event of the transaction with the string *"completed."* Now, consider the following events:

```
[11/Dec/2019:19:06:13] TSM backup_id=17 starting ...
[11/Dec/2019:19:34:43] TSM backup_id=17 %complete=50
[11/Dec/2019:20:01:19] TSM backup_id=17 %complete=75
[11/Dec/2019:20:33:34] TSM backup_id=17 i/o error.
```

Now we have a problem. The backup with backup_id 17 never completed. The backup failed with an error (i/o error). The `transaction` command will not group these events as transaction because the last event of the transaction will never be found. Transaction command will discard any event that doesn't belong to a transaction. How do we capture this type of unfinished transactions?

CHAPTER 4 GROUPING AND CORRELATING

First thing we need to instruct the transaction command is to keep the evicted transactions. Second, utilize the field *closed_txn* that the transaction command creates. This field is set to 1 for closed transactions and 0 for the evicted transactions. Rewriting our SPL

```
index="main" sourcetype="backup:log"
| transaction backup_id startswith="starting" endswith="completed"
keepevicted=true
```

The transaction command now has added an option keepevicted=true. The result will look like what is shown in Figure 4-5.

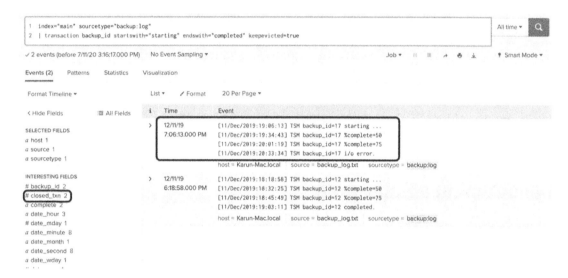

Figure 4-5. *Keeping the evicted transactions with the transaction command*

To exactly locate the evicted transactions, simply search for closed_txn=0:

```
index="main" sourcetype="backup:log"
| transaction backup_id startswith="starting" endswith="completed"
keepevicted=true
| search closed_txn = 0
```

The query simply filters for closed_txn=0. Any transaction with the field *closed_txn* set to 0 will be retrieved. In addition, the *duration* field is available for evicted transactions as well. This can be handy if you want to know how long the failed transaction took before failing.

125

CHAPTER 4 GROUPING AND CORRELATING

That sums up our discussion on the `transaction` command. There is one caveat with transaction command though. It is very resource intensive, and depending on the data set size, it can be slow to process. Keep that in mind when you utilize the command. We'll now move on to another important feature of SPL that helps in grouping and correlating events – *subsearches*.

Subsearches

SPL provides a way to run a search within a search. This is achieved by using *subsearches*. The results of a subsearch are returned to the main search where it can be utilized for further processing. Subsearches are often used to provide filtering for the main search.

The most common use of subsearches is to search for a particular piece of information that can be used as an argument to the main search. This particular piece of information is often dynamic and cannot be predetermined. When used with the `append` command, subsearches are also used to combine the results with main search. The term *inner search* and *subsearch* can be used interchangeably. Similarly, the term *outer search* and *main search* can be used interchangeably.

Constructing a Subsearch

Subsearches must be enclosed in square brackets. They typically begin with the command `search`. Syntax wise, a subsearch is no different than a normal search. An important point to note is subsearches are always executed first, before the main search.

Note Subsearches must begin with a generating command. Generating commands are those that generate events from one or more indexes without transforming them. Examples of generating commands include `search`, `tstats`, `inputlookup,` and `dbinspect`.

Let us take a look at how it works. For example, let's say you want to examine all the log events for the least sold product in the past two days. The search could be designed as follows:

CHAPTER 4 GROUPING AND CORRELATING

- The inner search (subsearch) finds the least sold product.

- The outer search (main search) retrieves all events for the least sold product (found by the inner search in the previous step):

```
index=main sourcetype=access_combined_wcookie
    [ search index=main sourcetype="access_combined_wcookie"
      action=purchase
    | stats count by productId
    | sort 1 count
    | fields productId ]
```

The inner search is enclosed in square brackets and begins with the command search. The search finds the least sold product by retrieving all the events with action=purchase and using stats and sort command to find the least sold product. The argument 1 in the sort command limits the output to one result. Finally, the fields command is used to just output the *productId* as we don't need the count. Let's say the least sold product was SF-BVS-01. The outer search, in effect, will behave like in the following:

```
index=main sourcetype=access_combined_wcookie productId="SF-BVS-01"
```

Let us consider another example. Splunk internal logs record the activities of the *splunkd* process. The events have a field named *component* that indicates the subsystem of splunkd. If you want to find the host with the greatest number of warning messages and break down the top five components that have the greatest number of errors for that host, you would construct the SPL as follows:

First, find the top host with the greatest number of warning messages:

```
index=_internal log_level=WARN
| top 1 host
| fields host
```

The preceding SPL will be the subsearch. The structure of the main search will look like the following:

```
index=_internal  log_level=WARN  << subsearch >>
| top 5 component
```

127

In this SPL the place holder << subsearch >> is where the subsearch is used. The main search calculates the top five components. The full SPL looks like the following:

```
index=_internal  log_level=WARN  [search index=_internal log_level=WARN
| top 1 host
| fields host]
| top 5 component
```

The result will look like the following:

```
Component                count      percent
---------------------------------------------
SearchAssistant          8122       84.104794
MessagesManager          627        6.492700
PipelineComponent        496        5.136171
SavedSplunker            106        1.097649
LocalAppsAdminHandler    52         0.538470
```

The subsearches can also be nested, which may be useful depending on the use cases you may have.

Problems with Subsearches

There are two pitfalls with using subsearches. For this reason, subsearches should be used only when it is absolutely necessary. In many situations, using `stats` combined with `eval` can do the job. Here are the two issues with subsearches.

First, there is a limit in the number of results a subsearch can handle. By default, the limit is 10,000. After this limit, the results are truncated, so there is a potential for the outer search to be inaccurate.

Second, there is a time limit on how long a subsearch can run. By default, this limit is 60 seconds. If a subsearch runs after this time, the search is finalized, and only the events found thus far are returned to the outer search. This can lead to partial results being sent back to the outer search.

The preceding two limits can be increased using *limits.conf* on the Splunk server side. But this comes with the cost of increased memory usage. Further, there is always the potential to go above the limit depending on the volume of data you are dealing with.

In addition, there is potentially another pitfall with subsearches. This has to do with time range. By default, the time range you specify in the time picker will be used for both outer search and inner search. When the outer search is run at real time though, the inner search's time limit is set to *All time*. It is not recommended to run the inner search with *All time* because it can potentially hit the time limit of 60 seconds and produce partial results. One way to work around this issue is to explicitly specify the time range for the inner search using the time modifiers `earliest` and `latest`. Note that **when the time range is specified explicitly using the time modifiers, they are effective only for the block of search they are affecting**. For example, the following search uses two different time ranges for outer and inner searches:

```
index=_internal log_level=WARN earliest=-1h@h latest=@h
    [ search index=_internal log_level=WARN earliest=-2d@d latest=now
    | top 1 host
    | fields host]
| top 5 component
```

The inner search runs with the time frame of the past two days whereas outer search runs with the time frame past hour. First, we find the top host with warning messages based on the past two days' worth of logs. Then, we retrieve the warning logs for that host for the past hour to review.

Now, we'll proceed to take a look at `join`, which utilizes subsearches for combining two result sets.

Join

In SQL (Structured Query Language), joins are an invaluable piece of functionality that can be used to combine data from two or more tables based on one or more related columns. SPL provides its own version of join that you can use to combine the results of a subsearch with results of a main search using one or more join fields.

CHAPTER 4 GROUPING AND CORRELATING

Constructing a Join

Let us consider an example. The following events are under sourcetype *backup:log* which represent the progress of a backup job:

```
[11/Dec/2019:18:18:58] TSM backup_id=12 starting ...
[11/Dec/2019:18:32:25] TSM backup_id=12 %complete=50
[11/Dec/2019:18:45:49] TSM backup_id=12 %complete=75
[11/Dec/2019:19:03:11] TSM backup_id=12 completed.
```

The *backup_id* in the preceding events represents the unique job identifier. Now, consider the following events under sourcetype *backup:log_details* which provide additional details about a given backup job:

```
[11/Dec/2019:18:18:58] TSM backup_id=12 user=tsm_admin numberof_files=28594 estimated_size_mb=5087 type=incremental policy=daily_incremental_prod_web storage=disk
[11/Dec/2019:19:06:13] TSM backup_id=17 user=tsm_admin numberof_files=458750 estimated_size_mb=308709 type=full policy=weekly_full_prod_web storage=disk
```

Notice how the *backup_id* in both the log files can be used to link the data. For example, if I want to find out details about backup_id 12, I can query the sourcetype *backup:log_details*. And if I want to find out whether backup id 12 completed or not, I can query sourcetype *backup:log*. But if I want to combine the results from the two sourcetypes, can I create a coherent report for all successful backup jobs?

You can use the `join` command to combine the results of a main search and a subsearch on a join field. The join field acts like the constraint using which the results are joined. In our example, we can use *backup_id* as the join field. The SPL query can be written as follows:

```
index=main sourcetype=backup:log completed
| join backup_id
    [search index="main" sourcetype=backup:log_details]
| table _time,backup_id,estimated_size_mb,type,user
```

The main search looks for all completed backup jobs. The results are piped into the join command which uses the field *backup_id* as the join field. The subsearch retrieves the backup log details. The *join* command combines the results of the main search and

CHAPTER 4   GROUPING AND CORRELATING

subsearch using the join field *backup_id*. In other words, events that have the same *backup_id* in both the results are combined. Finally, the `table` command prints the required fields. The result is shown in the following:

```
_time                backup_id    estimated_size_mb    type          user
-----------------------------------------------------------------
2019-12-11 19:03:11  12           5087                 incremental   tsm_admin
```

By default, `join` performs an *inner join* in which only the events that match on both main search and subsearch are kept in the final result. You can override this by using the join option *type=outer* in which all the events from the main search and only the events that are matched on the subsearch are kept. In place of *type=outer*, you can also use *type=left*. They produce the same result. The syntax is provided in the following:

```
| join type=outer [join field] [subsearch]
```

In addition, by default only the first matching result from a subsearch is combined with the main search. You can override this by providing `max=0` to match unlimited events from the subsearch.

Tip When using join, field values from subsearch will overwrite the fields of the main search. To prevent this, you can use the option `overwrite=false`.

One problem you may run into is the field names in main search and subsearch don't match all the times. For example, in main search the field of a transaction ID may be named as *transaction_id*, whereas in the subsearch, the field may be named as something else, such as *tran_id*. In these cases, you can use the `rename` command within the subsearch to match the main search. For example, in our previous example, if the backup ID is defined with the field name *bk_id* in the sourcetype *backup:log_details*, you can rename *bk_id* to *backup_id* to match the main search as follows:

```
index=main sourcetype=backup:log completed
| join backup_id
    [search index="main" sourcetype=backup:log_details
    | rename bk_id as backup_id]
| table _time,backup_id,estimated_size_mb,type,user
```

Problems with Join

Since `join` uses subsearches, the limitations of subsearches apply to joins as well. As you may recall, subsearches have restrictions on both the number of results it can return and the amount of time it can execute. When these limits are exceeded, partial results are returned to the main search which can result in inaccurate calculations.

The output limit for join can be increased by increasing *subsearch_maxout* in *limits.conf*. The default value for this limit is 50,000. Note that when subsearch is used without `join`, the limit is 10,000.

Wherever possible, using `stats` with `eval` should be tried in place of `join`. In addition to the subsearch limitations, joins are inherently less efficient than using `stats`.

We'll move on to discuss the `append` family of commands which provide yet another way to combine results of multiple searches.

Append, Appendcols, and Appendpipe

The `append` family of commands provides a way to combine results from dissimilar searches. For example, they can be used to combine search results with two different time frames. These commands utilize subsearches to retrieve results to be combined. An important behavior of these commands is that they do not work on the real-time searches. These commands only work on historical searches. Let us take a detailed look at each of these commands.

Append

The command `append` simply tacks on the results of the subsearch to the main search. The results of the subsearch can be raw events or some statistical output. Still the append command simply adds the result rows under the results of the main search. Let us consider an example. You would like to list the top two products sold yesterday using the Splunk tutorial data. You would also like to show the top two products of all time in the same results. One way to achieve this is to use the append command as shown in the following:

CHAPTER 4 GROUPING AND CORRELATING

```
index=main sourcetype="access_combined_wcookie" action=purchase earliest=-
1d@d latest=@d
| top limit=2 productId showperc=f
| eval timeperiod = "yesterday"
| append
    [ search index=main sourcetype="access_combined_wcookie"
      action=purchase earliest=1 latest=now
    | top limit=2 productId showperc=f
    | eval timeperiod = "all_time"]
```

That seems like a lot of code, so let us break this down. The main search finds the top two products sold yesterday. It uses *limit=2* option with top command to limit the number of rows returned to 2 and suppresses the output of percentage. It also adds a new field named *timeperiod* and sets the value to *yesterday*. The output of main search alone will yield the following result:

```
productId    count   timeperiod
-------------------------------
SC-MG-G10    75      yesterday
WC-SH-G04    71      yesterday
```

Note Splunk tutorial data may not produce any results for *yesterday* as it contains curated data from the past that may change upon new versions of Splunk. You may have to change the search time frame based on the copy of tutorial data you have. Alternatively, you can use data from your own environment instead of using the tutorial data.

The subsearch performs very similar search except two key differences. First, it sets the timeframe to All time by specifying *earliest=1*. Next, it creates a new field named timeperiod and sets the value to *all_time*. The subsearch alone will yield the following result:

```
productId    count   timeperiod
-------------------------------
WC-SH-G04    275     all_time
SC-MG-G10    273     all_time
```

133

CHAPTER 4 GROUPING AND CORRELATING

The combined search with *append* will yield the following result:

```
productId    count    timeperiod
-------------------------------
SC-MG-G10    75       yesterday
WC-SH-G04    71       yesterday
WC-SH-G04    275      all_time
SC-MG-G10    273      all_time
```

Let us consider another example. You want to list the number of 500 series HTTP errors that occurred in the last hour and compare this with the previous hour (these errors indicate internal server errors and are fatal server-side errors). The following SPL can do the job:

```
index=main sourcetype="access_combined_wcookie" status=5* earliest=-1h@h latest=@h
| stats count AS "LastHourErrors" by status
| append
    [ search index=main sourcetype="access_combined_wcookie"
      status=5*  earliest=-2h@h latest=@h
    | stats count AS "PreviousHourErrors" by status]
```

The main search grabs the HTTP statuses during the past hour that start with 5 in order to cover codes such as 500,503, and so on. It uses `stats` command to list the number of errors in the field *LastHourErrors* split by the status code. The subsearch does similar thing but with two differences. First, it sets the time frame to the hour prior to the last hour. It also names the count field as *PreviousHourErrors*. The result is as shown in the following:

```
status    LastHourErrors    PreviousHourErrors
----------------------------------------------
500       5
503       6
505       3
500                         3
503                         5
505                         5
```

CHAPTER 4 GROUPING AND CORRELATING

As you can see, there are two columns, one for *LastHourErrors* and the other for *PreviousHourErrors*. As indicated before, append simply tacks on the results of subsearch at the end of the main search. While this is useful, it may not always be desired. For instance, when you visualize the preceding statistics in a column chart, the chart shown in Figure 4-6 emerges.

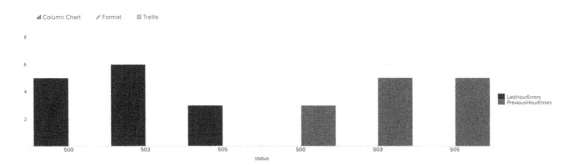

Figure 4-6. *Column chart resulting from append command*

As you can see, while the relevant data is included in the chart, it may be desired to overlay the count for a given status code. In other words, instead of tacking on to the results of the main search, it may be beneficial to place the results from subsearch *alongside* the main search. This is easily achieved through the appendcols command which is what we'll be discussing next.

Appendcols

The command appendcols behaves similar to append with one major difference. It overlays the results with the main search instead of tacking on to it at the end. The first subsearch result is merged with the first main result, the second subset result is merged with the second main result, and so on. The following example shows appendcols in action:

```
index=main sourcetype="access_combined_wcookie" status=5* earliest=-1h@h
latest=@h
| stats count AS "LastHourErrors" by status
| appendcols
    [ search index=main sourcetype="access_combined_wcookie"
    status=5*  earliest=-2h@h latest=@h
    | stats count AS "PreviousHourErrors" by status]
```

135

The only change I made in the preceding SPL query compared to the previous example is I replaced `append` with `appendcols`. The result is shown in the following:

```
Status     LastHourErrors    PreviousHourErrors
------------------------------------------------
500        5                 3
503        6                 5
505        3                 5
```

As you can see, the values are overlaid. This makes it easy to comprehend the data. The corresponding visualization in stacked column chart looks like Figure 4-7.

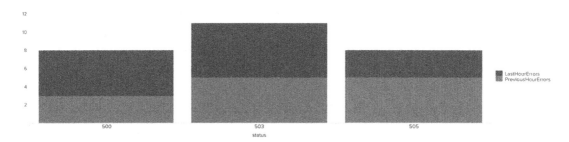

Figure 4-7. Column chart resulting from append command

There is a caveat with using `appendcols` though. When a field that is present in main search has no value in subsearch, `appendcols` ends up pushing up the data. This is because `appendcols` merges the first result of the subsearch with the first result of main search and repeats the same step for subsequent subsearch results. So, use `appendcols` with caution.

Appendpipe

The command `appendpipe` is the last command in the append family of commands. The major difference with `appendpipe` is that the subsearch is not executed first. The subsearch is executed when the search reaches the `appendpipe` instruction. The most common use of `appendpipe` is to calculate subtotals in a statistical output. Let us consider the following example:

```
index=main sourcetype="access_combined_wcookie" categoryId IN
("ARCADE","TEE","SPORTS")
| stats count by categoryId,productId
```

The SPL query generates a table that breaks down the traffic (count) for the three categories ARCADE, TEE, and SPORTS by every unique combination of *categoryId* and *productId*. The output looks like in the following:

```
categoryId      productId       count
-----------------------------------
ARCADE          BS-AG-G09       543
ARCADE          FI-AG-G08       599
ARCADE          MB-AG-G07       781
SPORTS          CU-PG-G06       555
TEE             MB-AG-T01       801
TEE             WC-SH-T02       619
```

Now, if you want to insert the subtotal for each category, you can easily do so with the following *appendpipe* query:

```
index=main sourcetype="access_combined_wcookie" categoryId IN
("ARCADE","TEE","SPORTS")
| stats count by categoryId,productId
| appendpipe
    [ stats sum(count) AS count by categoryId
    | eval productId = "TOTAL OF ALL PRODUCTS" ]
| sort categoryId
```

The preceding SPL pipes the result of the main search to appendpipe. Notice that the subsearch does not have to begin with a generating command such as search. This is because the subsearch is not run first. The main search runs first, and the subsearch is simply run in a subpipeline. The subsearch calculates the sum of the count by each *category id*. Notice also that in place of *productId* field, we add a descriptive name such as "TOTAL OF ALL PRODUCTS". Finally, we sort the results by *categoryId*. (If you don't sort, the total line shows up at the end of the main search results.) The output looks like the following:

```
categoryId      productId               count
---------------------------------------------
ARCADE          BS-AG-G09               543
ARCADE          FI-AG-G08               599
ARCADE          MB-AG-G07               781
```

ARCADE	**TOTAL OF ALL PRODUCTS**	**1923**
SPORTS	CU-PG-G06	555
SPORTS	**TOTAL OF ALL PRODUCTS**	**555**
TEE	MB-AG-T01	801
TEE	WC-SH-T02	619
TEE	**TOTAL OF ALL PRODUCTS**	**1420**

As you can see, the subtotal of each category has been added to the results.

Key Takeaways

We looked at important grouping and correlating commands in this chapter. As you go about using Splunk for your IT operations and security use cases, you will inevitably run into situations where you need to group and correlate events from multiple sources in a meaningful way. Mastering the commands discussed in this chapter will greatly help. Here are the key takeaways from this chapter:

1. You can use the `transaction` command to group conceptually related events, based on the constraints you provide.

2. Use field values and/or startswith/endswith options to narrow down your transactions.

3. The fields *duration* and *eventcount* are automatically created by transaction command that can be very useful for statistical calculations.

4. You can find unfinished transactions by using *keepevicted=true* and checking *closed_txn* field.

5. The `transaction` command has performance impact. So, use cautiously. In many cases, you can construct equivalent searches with `stats` and `eval` instead.

6. The `join` command lets you combine results from multiple sources based on a join field. Default mode of join is inner join.

7. The command append lets you tack on the results of a subsearch into the main search.

8. The command `appendcols` is similar to append but overlays the column data.

9. Use `appendpipe` when you need to process the commands in a subpipeline, most commonly used for creating subtotals.

10. Subsearches have inherent default limits on the number of results (10,000) and the amount of time (60 seconds).

You have come a long way in learning the basic building blocks of SPL. By combining the power of `stats`, `eval`, `timechart,` and grouping commands, there is only a very little you can't do with Splunk. In the next chapter, we'll take a small detour and take a deeper look at *fields*. Fields are integral part of machine data and crucial to effective use of SPL. Your ability to make the most out of SPL will depend on the availability of useful fields. We'll learn how to make use of the automatically extracted fields and discuss how to extract your own fields.

CHAPTER 5

Working with Fields

With over 140 commands and associated functions, SPL provides unprecedented flexibility to search and analyze massive amounts of unstructured data. Majority of the SPL commands use *fields*, and some of them *require* fields. In order to fully utilize the power of SPL, you should be able to represent your data in terms of *fields*. Fields are searchable key-value pairs in your data. They are the building blocks of SPL. As a Splunk user, you will inevitably run into situations where you need to first extract fields from your data in order to produce informative reports. In this chapter, we'll learn about fields and how to extract them. We'll first study the importance of fields in SPL; then we'll move on to learning about automatically extracted fields. We'll dive deep into manually extracting fields. We'll cover the command rex in detail with plenty of examples. Finally, we'll learn some important SPL commands that make use of fields, such as sort and dedup. By the end of this chapter, you will have enough knowledge to create and make use of fields in Splunk.

Why Learn About Fields?

In almost all examples in this book, we've been using fields. In your day-to-day interaction with Splunk, it won't be any different – you will be using fields in almost every search. Let's take a look at the significance of learning about fields.

Tailored Searches

Whereas you can use SPL to simply search for any string or phrase, the real power of SPL is its ability to tailor the searches using fields. For example, in order to find the events with HTTP status code 500, you could use the following SPL:

500

CHAPTER 5   WORKING WITH FIELDS

The SPL query contains just one string *500*. This is a perfectly legal SPL query. Strictly speaking, Splunk implicitly adds the search command at the beginning. There is nothing wrong with the SPL query. When executed, Splunk will search all the *default indexes* associated with the user (who is executing the query) for the time range selected in the time picker. When it finds events matching *500*, it retrieves the data and displays the results in the search interface. See Figure 5-1.

Figure 5-1. *Searching for a literal string in Splunk*

Note   When index is not explicitly specified in the SPL query, the default indexes are searched. Default indexes are configured by Splunk administrators for a given Splunk role. A user belongs to at least one Splunk role.

But there are many problems with this search. First, the string *500* is very generic, and it will match many strings other than the HTTP status code. For example, if the response time of the request (which is logged as the last field in the access log event) happens to be *500*, that event will be retrieved as well. Likewise, if a JSESSIONID has the string *500* in it, that event will be retrieved as well.

Second, the search has to search through the entire event until it locates all the matching strings in an event. A wasted effort as all we are interested is just the HTTP status code.

Finally, the search is going to be very inefficient because it does not do any filtering. For example, if we had specified index=main and sourcetype=access_combined_wcookie, the search would have been much more efficient as it greatly reduces the data it needs to sift through. Let us rewrite the SPL query as follows:

index=main sourcetype=access_combined_wcookie status=500

This is a much more accurate and efficient search as it focuses on just the status field and applies two additional filtering – index and sourcetype. More importantly, it paves the way for tailoring your searches in a variety of ways. For example, if you want to find all status codes that start with 5

index=main sourcetype=access_combined_wcookie **status=5***

Using the wildcard *, we are able to easily retrieve all HTTP status codes that begin with 5. It will match statuses *500,501,502,* and so on. And it gets even better. If we need to locate all the HTTP statuses between *300* and *599,* we can write the following SPL:

index=main sourcetype=access_combined_wcookie **status >= 300 AND status <= 599**

When it comes to tailoring searches, fields open a world of possibilities with eval functions. For example, if you want to categorize the HTTP status codes as *Client Error* and *Server Error,* you can easily do so using the following eval statement using the status field:

index=main sourcetype="access_combined_wcookie"
| eval category = case(status >= 400 AND status < 500, "Client Error", status >= 500, "Server Error")

The SPL query uses eval's case statement to evaluate *status* and assign the result to a new field named *category*. If the status is between *400* and *500*, the field *category* is assigned the value *Client Error*. If the status is greater than *500*, it is assigned the value *Server Error*. You can then use the field *category* to tailor your search. For example, to report on the number of errors by category

index=main sourcetype="access_combined_wcookie"
| eval category = case(status >= 400 AND status < 500, "Client Error", status >= 500, "Server Error")
| **stats count by category**

CHAPTER 5 WORKING WITH FIELDS

Splunk produces the following result:

```
category            count
--------------------------
Client Error        3818
Server Error        1432
```

Insightful Charts

With fields, you can produce informative charts and reports with ease. The most common use of fields in creating reports and charts is the ability to split the results using the by clause. For example, consider the following SPL:

```
index=main sourcetype="access_combined_wcookie" action IN ("view","addtocart","purchase")
| timechart count
```

The preceding SPL query plots the count of events over time where the *action* is *view* or *addtocart* or *purchase*. Splunk produces the following statistical output:

```
_time               count
--------------------------
2019-12-04          502
2019-12-05          2456
2019-12-06          2512
2019-12-07          2445
2019-12-08          2277
2019-12-09          2363
2019-12-10          2336
2019-12-11          1980
```

You can also visualize the preceding info in a column chart. See Figure 5-2.

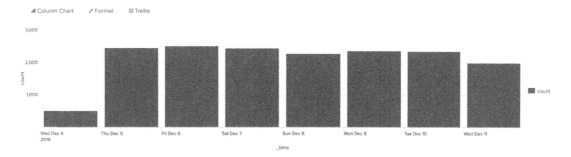

Figure 5-2. Visualizing timechart without a split-by field

Whereas the preceding chart is informative, it can impart even more useful data by splitting the data by the *action taken*. Consider the following SPL:

```
index=main sourcetype="access_combined_wcookie" action IN ("view","addtocart","purchase")
| timechart count BY action
```

The SPL query simply adds the split by clause by appending the statement by action. Splunk produces the following statistical output:

```
_time        addtocart    purchase    view
---------------------------------------------
2019-12-04   179          157         166
2019-12-05   837          783         836
2019-12-06   869          903         740
2019-12-07   813          829         803
2019-12-08   786          781         710
2019-12-09   799          793         771
2019-12-10   796          797         743
2019-12-11   664          694         622
```

145

CHAPTER 5 WORKING WITH FIELDS

And the corresponding column chart is shown in Figure 5-3.

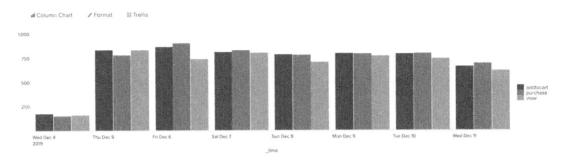

Figure 5-3. *Visualizing timechart with split-by field*

As you can see, splitting the count by action readily provides a visualization that breaks down the action taken.

Flexible Schema

One of the strengths of Splunk platform is its ability to apply a data schema when searching. This is known as *schema-on-read* or *schema-on-demand*. The traditional relational databases and many document-based data stores rely on a structure *before* the data is ingested. This is known as *schema-on-write*. In those systems, you have to painstakingly map out your fields before you ingest the data. And any changes to this schema result in incorrectly parsed data. With Splunk though, by using fields, you can apply any schema you want during search-time. This is a very powerful capability as it provides numerous possibilities to slice and dice the data. For example, consider the following events:

```
[11/Dec/2019:18:18:58] Exception occurred java.lang.InterruptedException
[11/Dec/2019:19:06:13] Exception thrown java.lang.RuntimeException
[11/Dec/2019:19:18:23] A fatal error occurred java.lang.
InterruptedException
[11/Dec/2019:19:35:50] java.lang.NullPointerException exception found
[11/Dec/2019:20:04:10] We may have caught a java.lang.NumberFormatException
[11/Dec/2019:21:33:22] java.lang.RuntimeException - program aborting
```

In these events, there are many java exceptions scattered around. They don't follow any particular pattern – that is, they don't appear in a certain place within an event. Mapping this out before ingesting the data into Splunk will be hard because we simply don't know where these exceptions will show up. In other words, creating fields before ingesting data is virtually impossible. With Splunk however, you can easily create a field named *exception* during search-time by using varieties of ways. For example, let me show you how you can use rex command to extract the *exception* field. Let's rewrite the SPL as shown in the following:

```
index="main" sourcetype="java:exception"
| rex "(?<exception>java\.[^\s]+)"
```

I have piped the events to rex command which uses the *regular expression* (?<exception>java\.[^\s]+) to create a new field named *exception*. The regular expression translates to any string that starts with java, until a space is found. We will cover rex command and regular expression in detail later in this chapter. The result of this SPL query is a new field named *exception* being created. To simply show the exceptions and their counts

```
index="main" sourcetype="java:exception"
| rex "(?<exception>java\.[^\s]+)"
| stats count by exception
```

Splunk produces the following result:

```
exception                         count
-----------------------------------------
java.lang.InterruptedException     2
java.lang.NullPointerException     1
java.lang.NumberFormatException    1
java.lang.RuntimeException         2
```

147

CHAPTER 5 WORKING WITH FIELDS

You can also visualize this in a pie chart. See Figure 5-4.

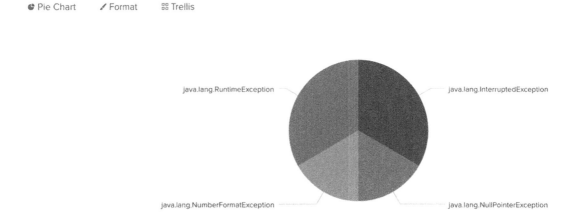

Figure 5-4. *A pie chart made possible by extracting fields during search-time*

As you can see, by using fields you are not dependent on a particular data schema. You can slice and dice the data in any way you want during search-time.

Index-Time vs. Search-Time Fields

In Splunk, fields can be created either during index-time or during search-time. As Splunk processes the incoming raw data during indexing, it can extract fields. This is called index-time field extraction. Many default fields such as *host*, *source,* and *sourcetypes* are extracted this way. You can also extract custom fields during index-time using *props.conf* and *transforms.conf* configuration files. As we have seen, Splunk can also extract fields during search-time. The rex command we example we saw earlier is an example of extracting fields during search-time. From SPL's perspective, whether a field is an index-time field or search-time field makes no difference. However, there are three major implications to consider with index-time fields.

First, because Splunk has to perform additional processing to extract fields during index-time, the overall indexing time is increased. In addition, once the data is indexed, any changes to the indexed fields will require reindexing the data.

Second, searches on the indexes can become slower because the index has been enlarged due to the index-time fields. In general, searching on a larger index takes longer to process.

Third, indexed fields must be written to disk and require more storage. The total additional storage burden is based on the number of additional indexed fields and the length of those fields' values.

As a general rule, perform the field extractions during search-time as it provides greater freedom and flexibility.

There is a benefit to using indexed fields. Metasearches, such `tstats` and `metadata`, operate only on indexed fields. `tstats` can be exponentially faster than `search,` and, in an environment where a field is used commonly but the data set spans terabytes or petabytes, this can be the preferred usage over search-time fields.

Automatically Extracted Fields

As Splunk processes the incoming data during the indexing phase, it automatically extracts many useful fields. These are also called *default fields*. These fields are kept in the index with the raw data. In addition, there are a few internal fields that Splunk creates during index-time. Internal field names being with _ (underscore), for example, _time. Further, Splunk also discovers new fields during search-time when search mode is set to either *smart* or *verbose*. This process is called *field discovery*. These three groups of automatically extracted fields (*default, internal,* and *discovered*) provide tremendous help in exploring your data. Let us learn about them in detail.

Default Fields

Splunk creates many default fields as it processes data during indexing. These fields can be very useful during search-time to filter your data. Let us take a look at these fields. Table 5-1 lists these fields and their descriptions.

CHAPTER 5 WORKING WITH FIELDS

Table 5-1. *Default fields made available by Splunk*

Field	Description
index	The index in which the data is indexed
host	The host from which the data was received
source	The source (generally the log file path) from which the data was received
sourcetype	The sourcetype used to parse the data Sourcetypes tell Splunk how to recognize the data
splunk_server	The Splunk server (indexer) that processed this data during indexing
linecount	Number of lines in the event. For example, a java stack trace can contain several lines but still indexed as one event
punct	Contains the punctuation pattern extracted during indexing
date_* fields	Provides additional granularity to the event timestamps. The fields include *date_hour, date_mday, date_minute, date_month, date_second, date_wday, date_year, date_zone*

Caution The *date_* fields are only available when the raw event contains a valid timestamp. Further, date_ fields are directly extracted from the raw event and do not take the time zone customizations you may have done into account.

An example of using default fields is shown in the following:

```
index="main" sourcetype=secure date_wday="sunday" "Failed password"
| stats count AS "Failed password attempts on Sundays" BY date_mday
```

The SPL query calculates the number of failed login attempts on Sundays and breaks it down by the date. Note the use of default fields *index, sourcetype, date_wday,* and *date_mday.*

Internal Fields

Splunk automatically creates many internal fields as it indexes the data. The names of internal fields begin with _ (underscore). Table 5-2 lists the internal fields.

Table 5-2. *Internal fields made available by Splunk*

Field	Description
_raw	The raw data of the original event
_time	Timestamp of the event in Unix time. The time is always stored in the index in UTC format and is translated to human-readable format during search-time
_indextime	The time at which the event was indexed, in Unix time. This field is hidden by default and will not be displayed unless renamed or used with `eval`

In addition to the preceding internal fields, there are also _cd which contains the address of an event within an index and _bkt which contains the *bucket ID*. Both of these fields are hidden by default. Use of these internal fields is not recommended for new users as they are used for internal reference only.

An example of SPL utilizing the internal index is shown in the following:

```
index=_internal sourcetype=splunkd
| eval latency = _indextime - _time
| convert ctime(_indextime) AS Indextime
| sort 4 -latency
| table Indextime,_time,latency
```

The SPL query calculates the *latency* between the timestamp of the event (_time) and the time at which it was indexed (_indextime). It also uses the command `convert` with `ctime` function to translate Unix time to human-readable time and stores the result in the *Indextime* field. It sorts the top four events using latency in descending order. Finally, it prints the results in a table. Notice the use of the internal fields *_time* and *_indextime*. Splunk produces the following result:

```
Indextime              _time                    latency
-------------------------------------------------------
07/22/2020 20:41:58    2020-07-22 20:41:55.314  2.686
07/22/2020 20:41:58    2020-07-22 20:41:55.315  2.685
07/22/2020 20:41:58    2020-07-22 20:41:55.315  2.685
07/22/2020 20:41:58    2020-07-22 20:41:55.315  2.685
```

We discuss sort command in detail later in this chapter.

CHAPTER 5   WORKING WITH FIELDS

Fields Extracted Through Field Discovery

Splunk automatically extracts key-value pairs present in the data. Key-value pairs are data represented in the format *key=value*, for example, *backup_id=20*. When Splunk sees events with key-value pairs, it automatically extracts them and makes them available for searching. In addition, you can configure Splunk to automatically extract fields when the data is of a structured format such as JSON. Field discovery is enabled only when the search mode is *smart* or *verbose*. It is not enabled in *fast* mode. Consider the following events:

```
[11/Dec/2019:19:03:11] TSM backup_id=12 completed.
[11/Dec/2019:19:06:13] TSM backup_id=17 starting ...
```

When these events are retrieved as the result of a search, Splunk automatically extracts the field *backup_id* since it is found in a valid key-value pair, *backup_id=12* and *backup_id=17*. You can see the automatically extracted fields under *interesting fields* in the fields sidebar. See Figure 5-5.

Figure 5-5. *Automatically extracted fields using field discovery*

Note that a field appears under interesting fields only if it is present in at least 20% of the events retrieved. But the field would still be present in the events, and you can use them in your searches.

CHAPTER 5   WORKING WITH FIELDS

When the data is of a structured format such as JSON, Splunk can be configured to automatically extract fields. By default, JSON fields will be automatically discovered by Splunk. Consider the following JSON event:

`{ "name":"George Washington", "role":"President", "country":"USA" }`

After the data is ingested into Splunk, when the event is retrieved using a search, Splunk automatically extracts the JSON fields. See Figure 5-6.

Figure 5-6. *Automatically extracted fields when data is of valid json format*

You can also see that Splunk highlights the syntax while displaying json data.

Whereas automatically extracted fields greatly aid in making use of your data, you will inevitably run into situations where you have to manually extract fields. In the next section, we'll learn about the various ways you can manually extract fields in Splunk.

Manually Extracting Fields

When the fields you need in your data are not automatically discovered and extracted by Splunk, you have to extract fields. This is also called custom field extraction. The custom field extractions must be applied to a *host, source,* or *sourcetype*. At first, manually extracting fields may seem like an intimidating task. In fact, in my experience, it is one of the most feared aspects of Splunk. The primary reason for this is the need to learn and understand *regular expressions*. If you want to extract fields from your data, you must learn to use regular expressions. While Splunk provides some aid in this area, such

153

as *field extractor wizard*, ultimately learning to use regular expressions will become inevitable in your Splunk journey. The good news is that you only need to learn a portion of regular expressions to tackle majority of the field extraction scenarios. In this section, we'll cover the various ways Splunk provides to manually extract fields.

Using Field Extractor Wizard

The easiest way to manually extract fields is to use the field extractor wizard. The field extractor wizard automatically generates regular expressions based on a sample of data you provide. If you want, you can update the regular expression that is generated by the wizard. This method is suitable if you are not comfortable writing your own regular expressions. The wizard can also use delimiters such as commas or spaces in your data to extract fields. This can be useful if your data is structured, such as a CSV file. Let us see field extractor in action. Consider the following set of events from Splunk tutorial data (sourcetype=secure):

```
Thu Dec 11 2019 00:15:06 mailsv1 sshd[4907]: Failed password for invalid user irc from 194.8.74.23 port 1956 ssh2
Thu Dec 11 2019 00:15:06 mailsv1 sshd[3014]: Failed password for invalid user operator from 194.8.74.23 port 1491 ssh2
```

These events show the failed login attempts for invalid users. Naturally, the *username* is an important piece of information, and it is worth extracting this as its own field. However, because it is not a key-value pair, Splunk does not automatically extract this field. In order to manually extract the username as a custom field, you can use the field extractor wizard. First, display the search results that include the usernames you want to extract as follows:

```
index=main sourcetype="secure" "invalid user"
```

The SPL query simply searches the main index with sourcetype secure and retrieves all events that have the string *invalid user* in them. Figure 5-7 shows the results. Notice that username is not one of the fields available.

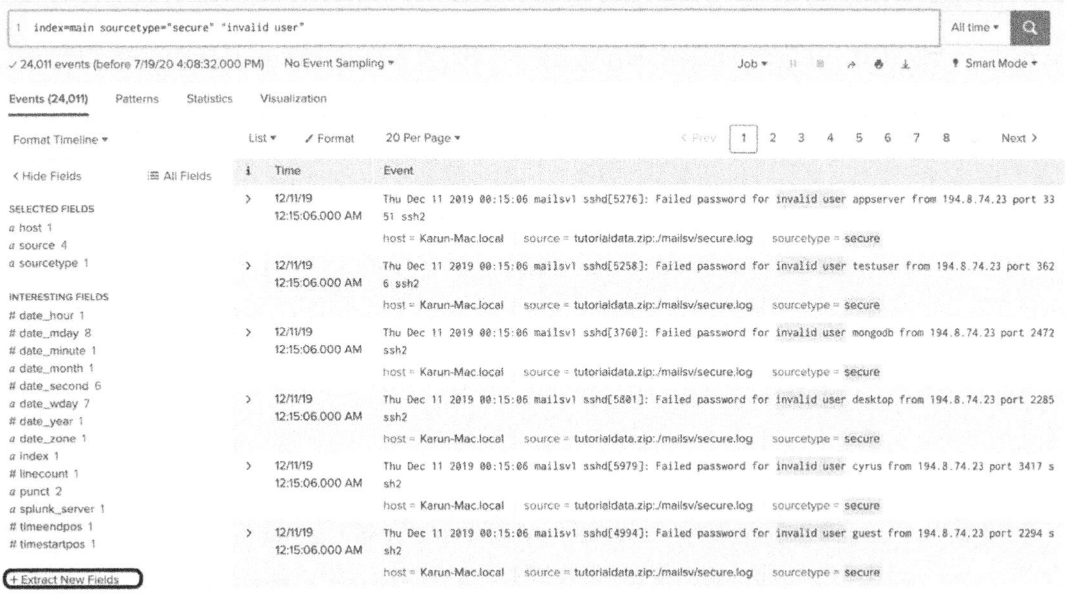

Figure 5-7. Displaying the targeted results before invoking the field extractor wizard

Click the *Extract New Fields* hyperlink toward the bottom of the fields sidebar. This opens the field extractor wizard. Choose a sample event that has the field you want to extract (username) from the list of events and click Next. See Figure 5-8.

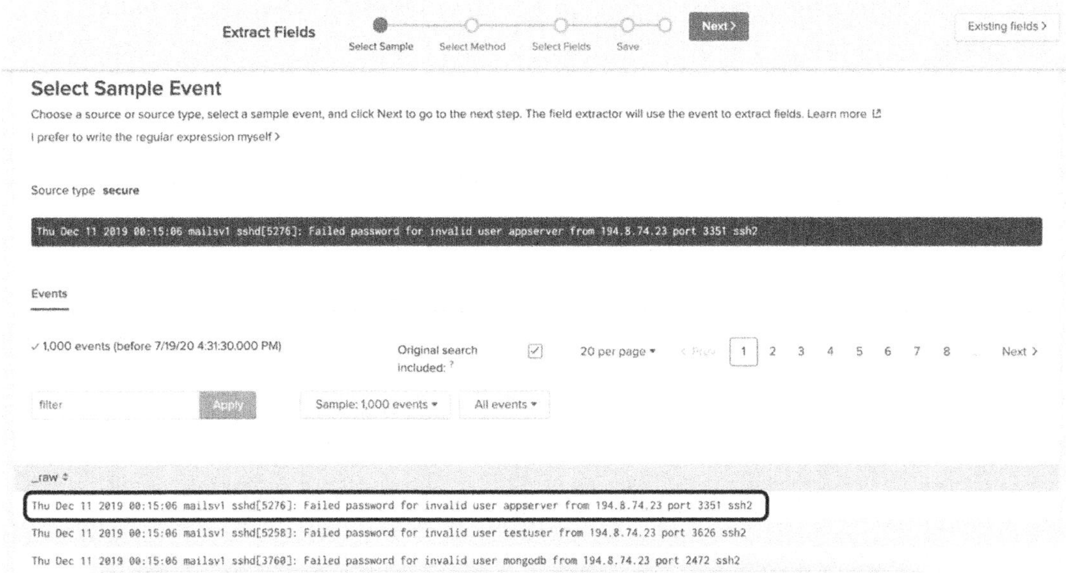

Figure 5-8. Selecting a sample event in the field extractor wizard

CHAPTER 5 WORKING WITH FIELDS

In the next screen, as shown in Figure 5-9, choose *Regular Expression* as the method to use. If your data is of structured data such as CSV, you can choose *delimiters* as the method. In our case, since the data is not structured, let's go with regular expression. Click Next to proceed.

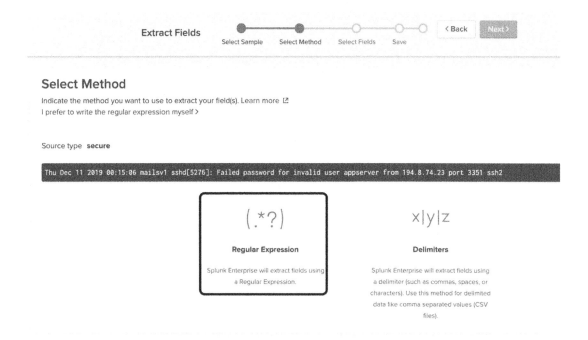

Figure 5-9. *Selecting Regular Expression as the method to be used for field extraction*

Now at the *Select Fields* screen, highlight the username (in this example, *appserver*). A pop-up is shown where you can enter a name for the field. As shown in Figure 5-10, enter *username* and click *Add Extraction*.

CHAPTER 5   WORKING WITH FIELDS

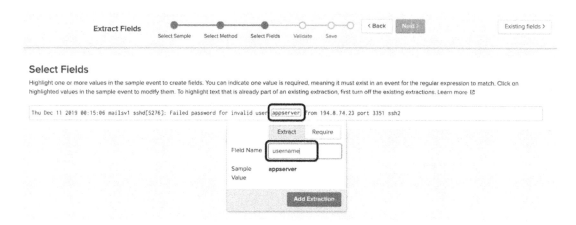

Figure 5-10. *Selecting the field to extractor in the field extractor wizard*

Now, Splunk goes to work and computes a regular expression that will best extract the field you selected. As shown in Figure 5-11, Splunk provides a preview of the extracted fields. If everything looks good at this screen, you can proceed by clicking Next. If you see any incorrect results, you can select additional events to the set of sample events. In our case, all the extracted fields look good. So, let's proceed by clicking Next.

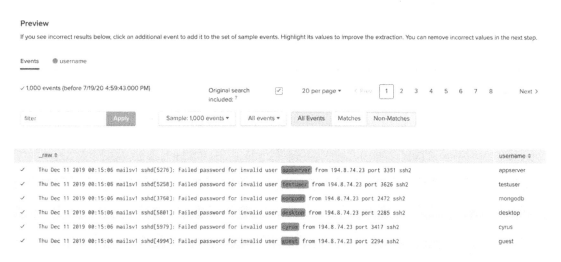

Figure 5-11. *Previewing the results in the field extractor wizard*

In the *Validate* screen, you have the option to remove any incorrectly extracted fields. In our example, we don't have any incorrectly extracted fields. In this screen, you also have the option to see the regular expression that Splunk generated for you. As shown in Figure 5-12, simply click Next.

CHAPTER 5 WORKING WITH FIELDS

Figure 5-12. Validating the results in the field extractor wizard

In the final save screen, you have the option to set the permissions for this field extraction. By leaving it as *Owner* as shown in Figure 5-13, this field extraction will remain private to you. If you want to share this field extraction with others, you have to choose *App*. For this example, let's leave it as Owner and click Finish.

Figure 5-13. Setting permissions for the field extraction in the field extractor wizard

CHAPTER 5   WORKING WITH FIELDS

A Success message appears as shown in Figure 5-14. Your field extraction is complete.

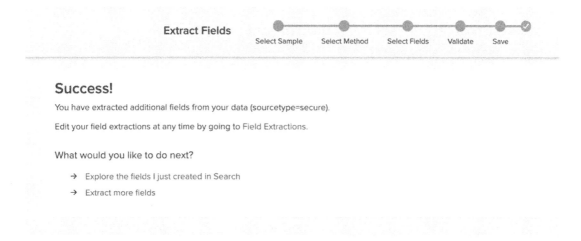

Figure 5-14. Completing the field extractor wizard

Go back to the Search & Reporting app and rerun the following search:

index=main sourcetype="secure" "invalid user"

Now, you can see the username field on the field sidebar as shown in Figure 5-14.

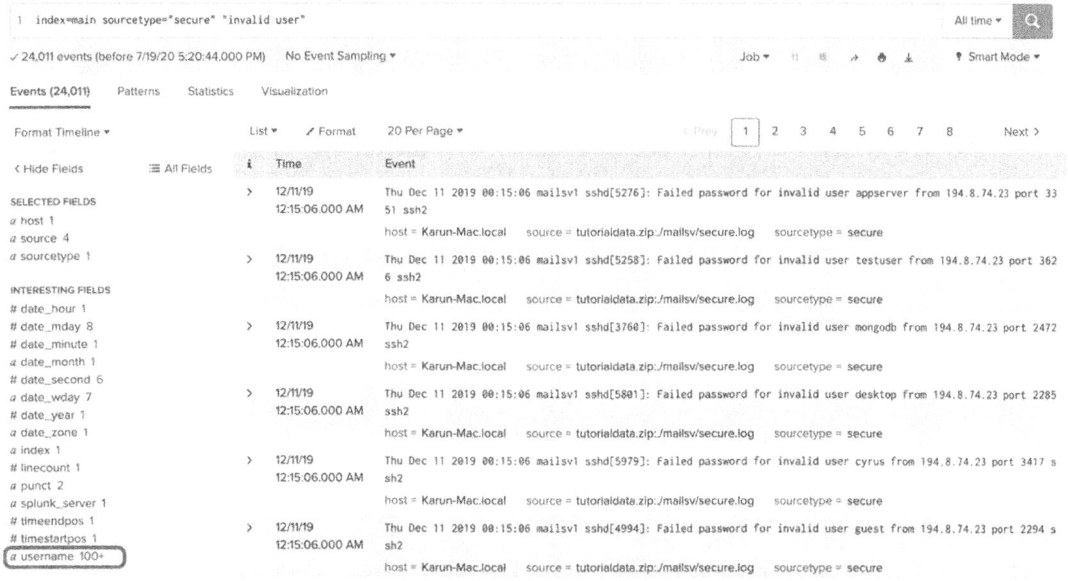

Figure 5-15. Validating the extracted field in search interface

159

Now that the field is extracted, you can tailor your searches to make use of the field. For example, to find the top ten invalid usernames, run the following SPL:

```
index=main sourcetype="secure" "invalid user"
| top 5 username showperc=f
```

By using `showperc=f`, I chose not to show the percentage to keep the report simple. Splunk produces the following result:

```
username              count
---------------------
administrator    1020
admin            938
operator         923
mailman          752
irc              644
```

Next, we'll take a look at how to use the field extractions menu to extract custom fields.

Using Field Extractions Menu

Whereas field extractor wizard is helpful for someone who is not comfortable with regular expressions, you will have more control over field extractions when you use regular expressions. One way you can extract custom fields is using *field extractions* menu under settings. You simply access the menu by navigating to *Settings* ➤ *Fields* ➤ *Field Extractions* as shown in Figures 5-16 and 5-17.

CHAPTER 5 WORKING WITH FIELDS

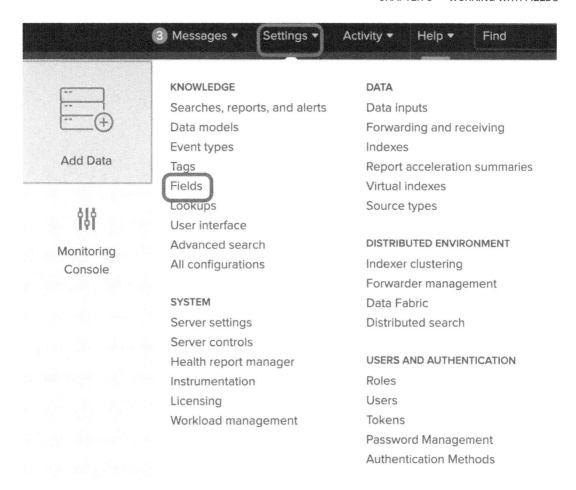

Figure 5-16. *Accessing the Field Extractions menu*

CHAPTER 5   WORKING WITH FIELDS

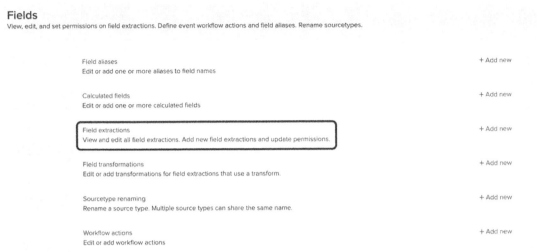

Figure 5-17. *Field extractions menu*

For illustration, consider the following events from the Splunk tutorial data:

```
Thu Dec 11 2019 00:15:06 mailsv1 sshd[5333]: Failed password for invalid user inet from 194.8.74.23 port 4564 ssh2
Thu Dec 11 2019 00:15:06 mailsv1 sshd[2605]: Failed password for invalid user itmadmin from 194.8.74.23 port 4692 ssh2
```

These events indicate failed login attempts. Note that the event has the user's IP address. This can be a very good piece of information for troubleshooting. Since the IP address is not in a key-value pair format, Splunk will not automatically extract the IP address as a field. In order to extract this as a field, you can create a new field extraction using the field extractions menu. Launch the field extractions menu from *Settings* ➤ *Fields* ➤ *Field extractions* ➤ *New Field Extraction*.

Enter the following information in the form:

Destination app: The Splunk app in which you want to create this field extraction. For this example, you can leave this as *Search*.

Name: The name for this field extraction. This is not the name of the field itself. That is specified as part of the regular expression.

Sourcetype: This is the sourcetype on which this field extraction will be applied. A field extraction must be applied to a *host, source,* or *sourcetype*. For this example, type *secure*.

CHAPTER 5 WORKING WITH FIELDS

Type: You can extract the fields in two ways. With *inline,* you specify the regular expression right on this page in the next step. You can also use *transform* which is yet another knowledge object that you configure outside of this field extraction. Using transforms allows you to reuse a piece of regular expression in multiple field extractions. For this example, choose *Inline.*

Extraction/transform: This is the actual regular expression to use. Splunk uses PCRE (Perl Compatible Regular Expression). By using named extraction of regular expressions, Splunk can create fields that match a regular expression pattern. We are going to go through a primer on regular expression in the next section. For now, in order to retrieve the IP address as a field named *userIP,* simply enter the following regular expression:

(?<userIP>(\d{1,3}\.){3}\d{1,3})

The regular expression captures the IP address pattern in a field named *userIP.* See Figure 5-18.

Add new
Fields » Field extractions » Add new

Destination app	search
Name *	userIP
Apply to	sourcetype named * secure
Type *	inline
Extraction/Transform *	(?<userIP>(\d{1,3}\.){3}\d{1,3})
	If the field extraction is inline, provide the regular expression. If the field extraction uses a transform, specify the transform name.

Cancel   Save

Figure 5-18. *Adding a new field extraction using the Settings menu*

Click *Save* to save the field extraction. Now when you search for the data, you can see the *userIP* field in the field sidebar. See Figure 5-19.

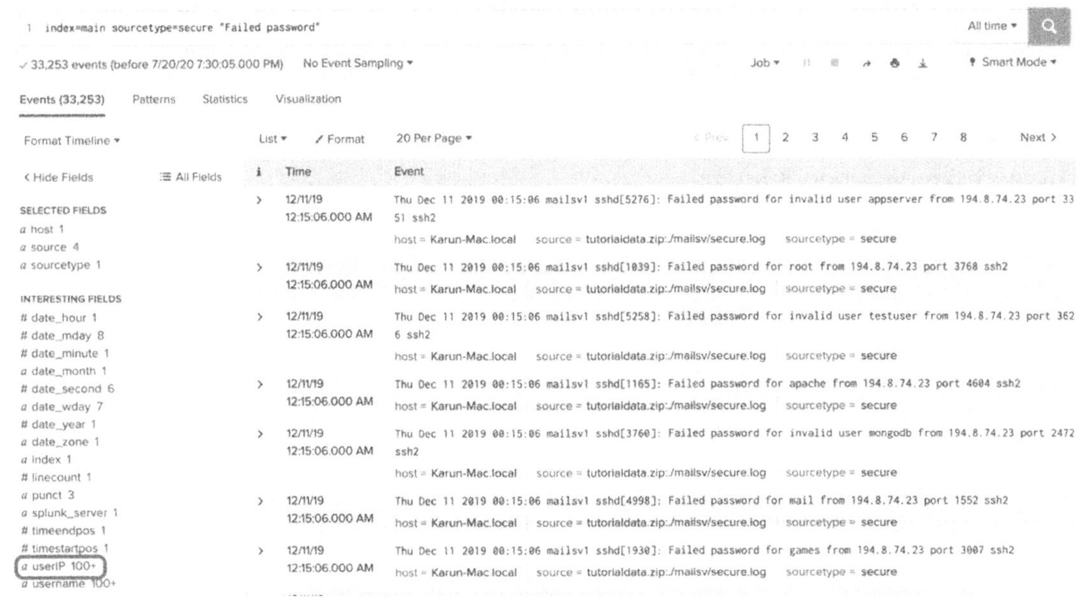

CHAPTER 5 WORKING WITH FIELDS

Figure 5-19. Extracted field show up in the fields sidebar

Once you have the *userIP* extracted, you can easily tailor your searches using that field. For example, to create a report that shows the top five countries based on the number of requests, you can use the following SPL query:

```
index=main sourcetype=secure "Failed password"
| iplocation userIP
| top Country showperc=f
```

The SPL query uses `iplocation` command which translates the IP address to a geolocation. Splunk produces the following output:

```
Country              count
---------------------------
United States        8691
China                5121
United Kingdom       2773
Russia               2457
South Korea          2118
```

CHAPTER 5 WORKING WITH FIELDS

As you have probably observed, extracting fields requires the working knowledge of regular expressions. In the next section, let's go through a quick primer on regular expressions.

A Primer on Regular Expressions

A regular expression is a sequence of characters that represent a pattern of text. There are many implementations of regular expressions available. Splunk utilizes PCRE (Perl Compatible Regular Expressions). Regular expressions are used in SPL commands like rex and regex and many configuration files like *props.conf* and *transforms.conf*. Most notably, regular expressions are heavily used in field extractions. Learning the full functionality of regular expressions is beyond the scope of this book. There are many books available that are dedicated to regular expressions. In this primer, we'll cover the most important aspects of regular expressions that pertain to Splunk.

Note Regular expression is also called regex.

A regular expression can contain *literal characters* and *metacharacters*. Literal characters represent the actual characters themselves. On the other hand, metacharacters have special meaning. The metacharacters are as follows:

[ ]\^$.|?*+(){}

One could say that learning regular expression is largely learning about metacharacters and their meaning. For example, consider the following regular expression:

an.

In the regular expression, the characters a and n are literal characters. The . (period) is a metacharacter and has a special meaning. It represents any character except a new line character. So, the regular expression can match the following strings:

ant

and

CHAPTER 5 WORKING WITH FIELDS

Similarly, the metacharacter \d represents a digit. Consider the following regular expression:

\d\d\d-\d\d-\d\d\d\d

The regular expression represents three digits followed immediately by a dash, followed immediately by two digits, followed immediately by a dash, followed immediately by four digits. They represent the format of a US social security number.

Table 5-3 shows the most common metacharacter patterns and their meanings.

Table 5-3. *Regular expression cheat sheet*

Metacharacter	Description	Example	Sample matching text
.	Match any character except new line character	s.n	Son sun
\d	Any digit	5\d\d	500 504
\w	Any word character (letter, number, or underscore)	\w\w\w\w	King num1
\D	Any nondigit	\D\D	Pi $$
\W	Any nonword character	\W\W	{} ()
?	Zero or one	cats?	Cat cats
+	One or more	\d+	9 358748
*	Zero or more	x*z	Xxxxz z
\	Escape character	\$*	$$$$ $

(*continued*)

Table 5-3. (*continued*)

Metacharacter	Description	Example	Sample matching text
{…}	Number of repetitions	\d{3}	458 *(Exactly 3 times)*
		\d{1,3}	10 *(1 to 3 times)*
		\d{2,}	12745 *(2 or more times)*
[…]	Character class	[aA]pe	Ape *(a or A)* ape
		[a-z]20	k20 *(range a-z)* b20
		[x-z4-6]	y5 *(range x-z followed by range4-6)* z4
		[^1-5]+	640 *(character that is not in the range 1-5)* 9
^	Beginning of the line	^Z	Zebra Zen
$	End of the line	\d$	Total 60 hi5
(?<var>…)	Named extraction	(?<ssn>\d{3}-\d{2}-\d{4})	111-11-1111 *(extracts the value in a variable named ssn)*
(?:…)	Logical grouping	(?:search|find)	search (matches *search* or *find*) find

Learning the metacharacters in the table will greatly improve your efficiency with field extraction. Note that there are many other advanced features of regular expressions that I did not cover. I personally use sites like https://regex101.com and https://www.rexegg.com/ to brush up my regular expression knowledge. We will use the regular expressions we have learned in the next section as we explore using the rex command to extract fields.

CHAPTER 5   WORKING WITH FIELDS

Using Rex Command

Rex extracts fields during search-time using the named extraction feature of regular expression. Usually, you use `rex` to try out the various regular expressions to use before setting up permanent field extraction using the Settings menu or using the configuration files. The basic syntax of `rex` command is as follows:

```
| rex <regular expression>
```

Let us learn to use rex using a few examples. Consider the following events:

```
2019-03-05 10:57:51.112   INFO org.apache.catalina.core.StandardEngine  :
Starting Servlet Engine: Apache Tomcat/7.0.52
2019-03-05 10:57:51.253   INFO o.s.web.context.ContextLoader            :
Root WebApplicationContext: initialization completed in 1358 ms
2019-03-05 10:57:51.698   WARN o.s.b.c.e.ServletRegistrationBean        :
Mapping servlet: 'dispatcherServlet' to
```

The events are from a hypothetical spring boot application log. Note that the severity of the log event is represented by the string right after the timestamp, such as INFO, WARN, and so on. Let's say you want to create an alert that keeps track of the number of WARN messages and triggers the alert when the number of WARN messages goes above a certain threshold. In order to achieve this, you would first want to extract the severity as a field. Since the severity is not present in a key-value pair format, Splunk will not automatically extract this field.

First, come up with the regular expression that matches the string you want to extract. Then, you can use this regular expression in `rex` command to extract the value as a field. The regular expression can be written as follows:

```
\d{3}\s+(?<log_level>(?:ERROR|INFO|WARN|DEBUG))
```

The regular expression matches a three-digit number followed by one or more spaces, followed by a capture group. The capture group uses the literal strings with the conditional function of OR (the pipe character). The regular expression will match any of the strings ERROR, WARN, INFO, DEBUG, and TRACE. For example, in the following line , "112. INFO" will be matched, and the string "INFO" will be captured in the field *log_level*:

```
2019-03-05 10:57:51.112   INFO org.apache.catalina.core.StandardEngine  :
Starting Servlet Engine: Apache Tomcat/7.0.52
```

CHAPTER 5 WORKING WITH FIELDS

Using the regular expression, the rex command can be written as follows:

...| rex "\d{3}\s+(?<log_level>(?:ERROR|INFO|WARN|DEBUG))"

The entire argument to rex command must be enclosed in double quotes. See Figure 5-20 that shows the result of the rex command.

Figure 5-20. *Using rex to extract fields*

Once the *log_level* field is extracted, it is easy to create a timechart using the following SPL:

```
index="main" sourcetype="spring:log"
| rex "\d{3}\s+(?<log_level>(?:ERROR|INFO|WARN|DEBUG))"
| search log_level = "WARN"
| timechart count AS "Number_of_WARN_messages"
```

The result of the SPL can be saved as an alert to send an email or page to engage the support team. Creating and configuring alerts is beyond the scope of this book.

169

CHAPTER 5 WORKING WITH FIELDS

Let us consider another example. The following event shows the time taken to initialize the application in milliseconds:

```
2019-03-05 10:57:51.253  INFO o.s.web.context.ContextLoader        :
Root WebApplicationContext: initialization completed in 1358 ms
```

Let's say you want to extract the time taken to initialize as a field. Once again, because the time taken is not in a key-value pair format, Splunk will not automatically extract. But with rex command, you can easily extract this by using a regular expression.

We see that the time taken is a sequence of numbers followed by space followed by the string *ms*. It is also preceded by the string "initialization completed in". The rex command can be written as follows:

```
...| rex "initialization\scompleted\sin\s(?<time_taken>\d+)\sms"
```

The rex command extracts the value \d+, which stands for one or more digits into the field *time_taken*. Note that the pattern must be followed by a space (\s) and the string ms. See Figure 5-21 to see the result produced by rex.

Figure 5-21. *Using rex to extract fields*

Once you have the time taken extracted as a field, it is easy to tailor your searches. For example, to find the events where time taken is more than 5 seconds (5000 milliseconds), the following SPL can help:

```
index="main" sourcetype="spring:log"
| rex "initialization\scompleted\sin\s(?<time_taken>\d+)\sms"
| search time_taken > 5000
```

Now that you have learned how to extract fields, let's take a look at some of the practical uses of fields.

Using Fields

Fields are an inseparable component of SPL. In this section, let's review some of the practical ways you can make use of the fields.

Filtering

In order to filter your search results, you can use the `search` and `where` commands. When the `search` command is the first command in the query, before the first pipe, it is used to retrieve events from index. Note that the `search` command is implied at the beginning of any search, and hence you do not need to specify it at the beginning of your search criteria. When the `search` command is not the first command in the pipeline, it is used to filter the results of the previous command in the pipeline.

When employing the `search` command to filter data using fields, you can compare the field values using Boolean and comparison operators. For example, to retrieve events that have the value of the field *bytes* greater than 3072

```
...| search bytes > 3072
```

The `where` command evaluates an expression and retains an event if the result is true. Otherwise, it discards the events. For example:

```
...| where (bytes/1024) > 3
```

The SPL query retrieves events where kilobytes *(bytes/1024)* is greater than 3 and discards all other events. Note that events that don't have the *bytes* field are also discarded. In addition to using an expression, where command can be used to compare two different fields. For example:

`...| where heap_usage > 0.9 * max_heap`

The SPL query retrieves evens where heap usage is more than 90% of the maximum heap. Both *heap_usage* and *max_heap* must be fields that exist in the event.

The expression used in the where clause can be any valid eval expression. For example, to retrieve the events where the IP address (*ip* field) falls within a CIDR (Classless Interdomain Routing) network, you can use the following where clause:

`...| where cidrmatch("10.0.0.0/23",ip)`

Another set of useful commands to filter is `head` and `tail`. The command `head` retrieves ten events from the top, and `tail` retrieves ten events from the bottom. You can specify the number of events you want to retrieve. For example:

```
index="main" sourcetype="spring:log"
| head 5
```

The SPL query retrieves five events from the top. Note that the event order is relative to the input order. In other words, `head` retrieves the *latest* events from the index.

To retrieve 20 events from the bottom (oldest events)

```
index="main" sourcetype="spring:log"
| tail 20
```

The events will be printed in reverse order with the oldest event printed first. You can also use `head` command with an `eval` expression. The events will be printed until the condition is true. For example, to retrieve events until the *log_level* changes from WARN, you would use the following SPL:

`...| head (log_level="WARN")`

The SPL query will retrieve first events until it reaches an event where *log_level* is NOT equal to WARN.

Sorting

You can use the `sort` command to sort the results by the specified fields. By default, `sort` orders result in ascending order. For example, to sort the results by the count field

```
index="main" sourcetype="access_combined_wcookie"
| stats count by action
| sort count
```

To sort in descending order, add the minus symbol as shown in the following:

```
...| sort -count
```

By default, `sort` limits the number of results returned to 10,000. To override this, specify 0 as shown in the following, which sets the number of results to unlimited:

```
...| sort 0 count
```

Note Using sort 0 can have negative performance impact due to the amount of data.

You can use more than one field to sort. If multiple fields are specified, sorting happens in order. Consider the following example:

```
...| sort -field1,+field2
```

The SPL query first sorts field1 in descending order and then sorts field2 in ascending order. The symbol + is optional.

Splunk sorts strings in lexicographical order. Few points to note about lexicographical order are given in the following:

- Numbers are sorted before letters.
- Uppercase letters are sorted before lowercase letters.
- Numbers are sorted by first digit of the number, for example, 1,10,198,204,65,9.

CHAPTER 5 WORKING WITH FIELDS

Deduping

In many occasions, you will want to remove duplicates from your data. The dedup can be used for those situations. By default, dedup command keeps only the first result of a unique value of the field specified. For example, to keep the first result of each unique *host*

```
...| dedup host
```

You can specify a count to be used for the number of results to keep. For example, to keep the first three results for each unique *sourcetype* value

```
...| dedup 3 sourcetype
```

You can also use multiple fields to dedup. For example, to keep the first two results for each unique combination of *host* and *source* values

```
...| dedup 2 host,source
```

Note that the results returned are usually the latest events in the index. You can override this by using the sortby option. For example, consider the following SPL:

```
index="main" sourcetype="access_combined_wcookie" categoryId != NULL
| stats count by categoryId,itemId
| dedup 2 categoryId sortby -count
```

The SPL query retains the top two *categoryId* after sorting the results by the number of items sold. Splunk produces the following result:

```
categoryId      itemId     count
--------------------------------
STRATEGY        EST-13     197
STRATEGY        EST-14     189
ARCADE          EST-16     115
ARCADE          EST-21     114
ACCESSORIES     EST-7      91
ACCESSORIES     EST-6      86
TEE             EST-27     85
TEE             EST-19     83
SHOOTER         EST-15     69
```

SHOOTER	EST-21	62
SIMULATION	EST-14	60
SIMULATION	EST-27	56
SPORTS	EST-16	37
SPORTS	EST-12	33

Key Takeaways

Your experience with Splunk will greatly depend on your ability to make use of fields from your machine data. You should try your best to get the fields automatically extracted by Splunk by ingesting structured files or by using key-value pairs. But when that is not possible, you should be comfortable enough to manually extract fields. Learning regular expression will greatly help in manually extracting fields. Here are the key takeaways from this chapter:

1. Fields are searchable key-value pairs in the machine data that can be either automatically or manually extracted.

2. Fields enable you to construct tailored searches, perform data transformation, and create informative charts.

3. Fields can be created at index-time or search-time.

4. Search-time field extraction is preferred. Index-time field extraction consumes more disk space.

5. You can use field extractor wizard, *Fields* menu, `rex` command, or configuration files (props.conf, transforms.conf) to extract fields.

6. `rex` command lets you extract fields on the fly by using SPL during search-time.

7. Developing expertise in regular expressions greatly helps with extracting fields.

8. Use the command `dedup` command to remove duplicates.

9. Use the `sort` command to sort the results. You can use `sort -field` to sort the field in descending order.

10. The commands `sort` and `dedup` are resource intensive, so use it with caution.

You have made tremendous progress in learning SPL. Until now, we've been using SPL against the indexed data. Splunk also provides a way to bring in external data to enrich your search results. It achieves this by employing *lookups*. In the next chapter, we'll learn all about lookups.

CHAPTER 6

Using Lookups

So far, we've been learning how to make use of the machine data that's already ingested in Splunk. SPL provides numerous commands to explore the data effectively. At times however, you may need to bring in external data into Splunk. For example, you may create a report of the top five user IDs that have the greatest number of login failures. The SPL might look like the following:

```
...| search status="denied" reason="incorrect password" | top 5 username
```

The result might look like the following:

```
username           count
-------------------------
jbarnes            1020
ksmith21           938
rmanroe            923
sjackson4          752
hbernstien         644
```

While the preceding information is helpful, it would be even clearer if we were able to add the full name and perhaps email address of these users. This can help, for instance, to send them email warnings. But more often than not, this kind of detailed information is not necessarily part of machine data. But it may be available in an external file or database. Assume you have a CSV (comma-separated value) file that has the following information:

```
username           fullname              email
----------------------------------------------------
jbarnes            James Barnes          j.barnes@acme.com
ksmith21           Kelly Smith           k.smith@acme.com
rmanroe            Rita Manroe           r.manroe@acme.com
sjackson4          Stephanie Jackson     sjackson@acme.com
hbernstien         Harry Bernstien       hberns@acme.com
```

CHAPTER 6   USING LOOKUPS

So, the problem is this: how can we make use of the data that is outside of what is already in the Splunk indexes. This is the problem lookups solve. Lookups enable you to enhance your search results by bringing in data from external sources.

In this chapter, we'll learn how to create and use Lookups. In addition, we'll also learn how to maintain the contents of the lookups.

Types of Lookups

Lookups enable you to add fields to your search results. These fields also appear on the *fields sidebar*. There are five types of lookups you can define in Splunk:

File-based

> You will upload a CSV file containing the data to be used as lookup.

External script

> You will use a python script or an executable binary file such as a c++ executable that is run during lookup, and the results are used for enriching the results.

KV store

> A KV store is a Splunk knowledge object that stores data in multidimensional key-value pairs. KV stores must be defined in *collections.conf* configuration file. You can use a KV store as a source for lookups.

Geospatial

> You can use geospatial lookups to retrieve geographical data that can be used to represent a region in the map.

DB lookup

> By using *DBConnect* Splunk add-on, you can create an external database-based lookup and use `dbxlookup` command to perform the lookup.

Among the five types of lookups, the most common type of lookup is the *file-based lookup*. So, we will discuss that in detail in this chapter.

File-Based Lookups

In a *file-based lookup*, you will first create a CSV file with the information you want to use as lookup. You will then upload this CSV file to Splunk as a lookup table file. Next, you will create a lookup definition that points to the lookup table file. Once a lookup definition is created, you can use the command lookup to query the lookup table.

The CSV file to be used as lookup table must contain only ASCII characters and no special formatting characters. If you are using Microsoft Excel, you can save the file in *MS-DOS comma-separated* format. The CSV file must have a header line. The column names of the header line become the fields in Splunk. The CSV file should contain at least two columns. The column names do not have to be the same as the field names in your events. However, in order to make use of the lookup, at least one of the column's values should match one of the fields in your events. This is also known as the *join field* or *match field*. In Figure 6-1, the highlighted field (username) is the join field that is present in both search results and in the lookup table.

Events		Lookup table		
username	count	**username**	fullname	email
jbarnes	1020	**jbarnes**	James Barnes	j.barnes@acme.com
ksmith21	938	**ksmith21**	Kelly Smith	k.smith@acme.com

Figure 6-1. *Comparing contents of events in index and the lookup table*

In general, avoid large lookup tables due to performance implications. Extremely large lookup tables (size over 100MB) can cause performance issues in a distributed search scenario. This is due to the fact that the search head must push the lookup tables to all search peers (indexers) during search. Distributed search is beyond the scope of this book. Let's take a look at how to create a CSV-based lookup table.

Creating a Lookup Table

We will use Splunk tutorial data's lookup table *prices.csv.zip* as example. This file can be downloaded from the following link:

```
https://docs.splunk.com/Documentation/Splunk/latest/SearchTutorial/
Systemrequirements#Download_the_tutorial_data_files
```

CHAPTER 6 USING LOOKUPS

The first three lines of the file are shown in the following for reference:

```
productId,product_name,price,sale_price,Code
DB-SG-G01,Mediocre Kingdoms,24.99,19.99,A
DC-SG-G02,Dream Crusher,39.99,24.99,B
```

As you can see, the first line is the header line that defines the column names. The Splunk tutorial data we've been using has the *productId* field in the events. But it does not contain the sale price of a product. This lookup table can be used to pull the sale price. In addition to sale price, you can pull other pertinent information such as product name. Download the prices.csv.zip to your computer and expand the zip file.

Uploading the Lookup Table File

You can use Splunk Web to upload the lookup table file. Navigate to *Settings* ➤ *Lookups* ➤ *Lookup table files*. Click *New Lookup Table File*. See Figure 6-2.

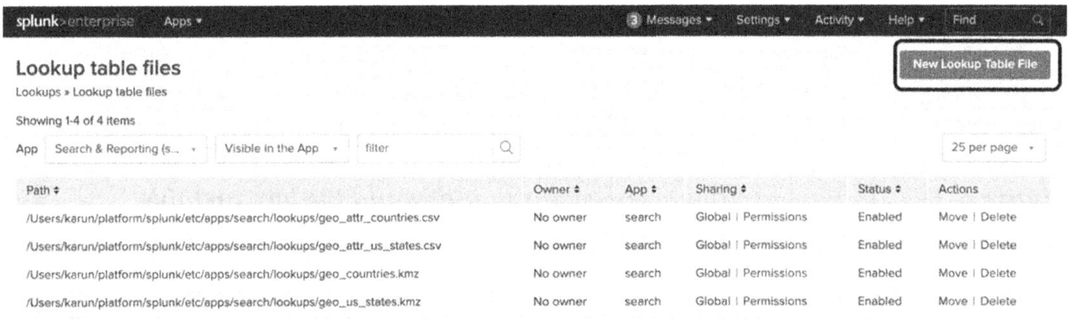

Figure 6-2. Creating a lookup table file

In the *Add new* screen, select *prices.csv* that you had downloaded and uncompressed earlier, and enter *prices.csv* in the *destination file name* field. As shown in Figure 6-3, click *Save*.

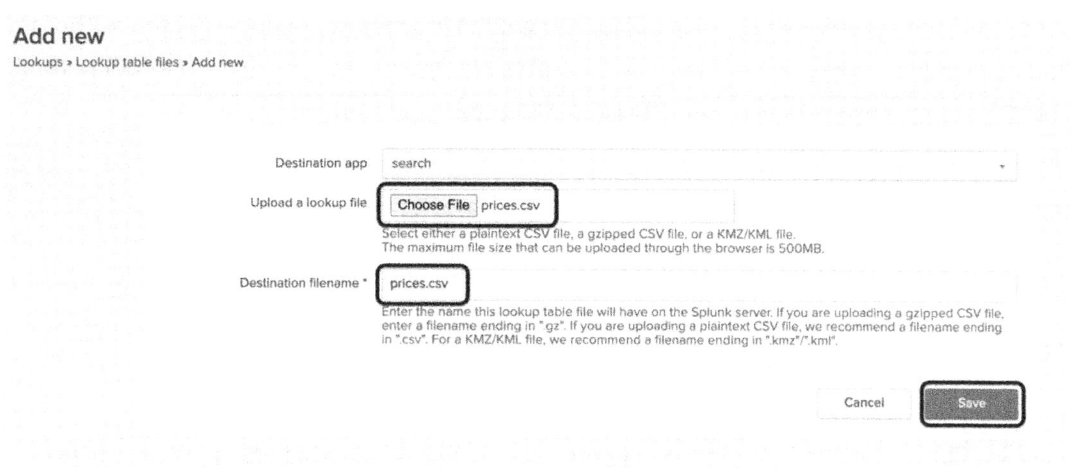

CHAPTER 6 USING LOOKUPS

Figure 6-3. Providing details to create a lookup table file

Next, create a lookup definition by navigating to *Settings* ➤ *Lookups* ➤ *Lookup definitions*. Click *New Lookup definition*. As shown in Figure 6-4, enter *prices.csv* for the *Name* and choose *prices.csv* from the drop-down for *Lookup file*, and then click *Save*.

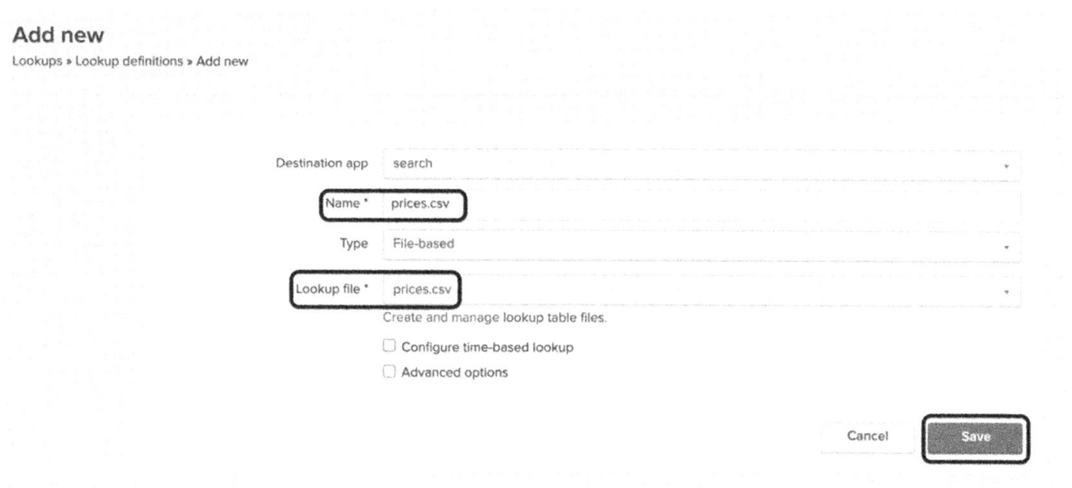

Figure 6-4. Providing details to create a lookup definition

You have now successfully uploaded the lookup table and created a lookup table definition. At this point, you can optionally set the permissions on the newly created lookup table. By default, the lookup table will be private to the creator. If you need to

CHAPTER 6   USING LOOKUPS

share this lookup table with other users, you will have to share the lookup table either in *app* or in *global* scope. We will not discuss the permissions and security in detail in this book. Now, you can proceed to verify the lookup table contents.

Verifying the Lookup Table Contents

Whenever you create a new lookup table, you must first verify the contents of the lookup table. You can easily do so with the `inputlookup` command:

```
| inputlookup prices.csv
```

The `inputlookup` command is a generating command that generates events from the lookup tables. It requires a leading pipe character and accepts the lookup table definition as the parameter. An excerpt of the output is shown in Figure 6-5.

Code	price	productId	product_name	sale_price
A	24.99	DB-SG-G01	Mediocre Kingdoms	19.99
B	39.99	DC-SG-G02	Dream Crusher	24.99
C	24.99	FS-SG-G03	Final Sequel	16.99
D	24.99	WC-SH-G04	World of Cheese	19.99
E	9.99	WC-SH-T02	World of Cheese Tee	6.99
F	4.99	PZ-SG-G05	Puppies vs. Zombies	1.99
G	19.99	CU-PG-G06	Curling 2014	16.99
H	39.99	MB-AG-G07	Manganiello Bros.	24.99
I	9.99	MB-AG-T01	Manganiello Bros. Tee	6.99
J	39.99	FI-AG-G08	Orvil the Wolverine	24.99

Figure 6-5. Verifying content of lookup table with inputlookup command

You can search for a particular piece of information by using the `search` command. For example, to retrieve the product ID and product names of products whose sale prices are more than $20

```
| inputlookup prices.csv
| search sale_price > 20
| table productId,product_name,sale_price
```

The SPL query filters the events from the lookup table by searching for `sale_price` > 20 and using the `table` command to output only the required fields. The output looks like the following:

```
productId          product_name         sale_price
-----------------------------------------------------
DC-SG-G02          Dream Crusher        24.99
MB-AG-G07          Manganiello Bros.    24.99
FI-AG-G08          Orvil the Wolverine  24.99
SF-BVS-G01         Grand Theft Scooter  21.99
SF-BVS-01          Pony Run             41.99
```

Now that you have uploaded and verified the contents of lookup table, it's time to start using it.

Using Lookups

In order to use the lookup tables, you use the `lookup` command. This is when you would match one or more events from your events to the data in the lookup table. You can retrieve all the columns from the lookup table or be selective using the OUTPUT option. If a column in the lookup table already exists in your events as a field, they are overwritten by default. You can override this behavior by using OUTPUTNEW option. When OUTPUTNEW is used, lookup is not performed for events in which the output fields already exist. Let us look at the lookup command in detail.

The Lookup Command

The basic syntax of the lookup table is as follows:

```
...| lookup <lookup-table> <lookup-field> AS <event-field>
```

The *lookup-table* is the name of the lookup table, *lookup-field* is a column name in the lookup table, and *event-field* is the field name in your search results that you want to match with the column name in the lookup table. If the field name is the same in both lookup table and search results, you don't need to specify the *AS* clause. By default, the

CHAPTER 6   USING LOOKUPS

lookup command will retrieve all columns from the lookup table except the matching column. For example, consider the following SPL query that retrieves the top five best-selling product of all time using the Splunk tutorial data:

```
index=main sourcetype=access_combined_wcookie action=purchase
| top 3 productId showperc=f
```

Splunk produces the following result:

```
productId    count
---------------
WC-SH-G04    275
SC-MG-G10    273
DB-SG-G01    266
```

The SPL query first filters data by using the constraint `action=purchase`. It then retrieves the top three *productIds*. It disables showing the percentage to simplify the report. Now, if you want to add the fields from the lookup table, for matching *productId*, you would simply use the following SPL query:

```
index=main sourcetype=access_combined_wcookie action=purchase
| top 3 productId showperc=f
| lookup prices.csv productId
```

The SPL query uses the *productId* as the match field. Splunk produces the following result:

```
productId    count  Code  price   product_name        sale_price
----------------------------------------------------------------
WC-SH-G04    275    D     24.99   World of Cheese     19.99
SC-MG-G10    273    L     19.99   SIM Cubicle         16.99
DB-SG-G01    266    A     24.99   Mediocre Kingdoms   19.99
```

As you can see, Splunk added additional fields to the search results. It matched the *productId* from the search results to the *productId* column in the lookup table. This is the power of lookup, to enrich your search results using data from external sources.

CHAPTER 6 USING LOOKUPS

If the field names in the lookup table and the search results are different, you must use the AS clause to perform the mapping. For example, if the field name in the lookup table is *productNumber* instead of *productId*, the lookup command must be modified as follows:

...| lookup prices.csv productNumber AS productId

Note Lookup table field values are case sensitive. The field values in the events and column values in the lookup table must exactly match. If you want to perform case-insensitive lookup, you must create a *lookup definition* based on the lookup table file. Lookup definitions are Splunk knowledge objects, and they are not covered in this book, but you can refer to the Splunk documentation at https://docs.splunk.com/Documentation/Splunk/latest/Knowledge/Usefieldlookupstoaddinformationtoyourevents.

Instead of adding all the columns to your search results as fields, you can pick the columns you want to add by using the OUTPUT option. For example, if you just want to add *product_name* and *sale_price*

```
index=main sourcetype=access_combined_wcookie action=purchase
| top 3 productId showperc=f
| lookup prices.csv productId AS productId OUTPUT product_name,sale_price
```

Splunk produces the following result:

```
productId   count   product_name        sale_price
---------------------------------------------------
WC-SH-G04   275     World of Cheese     19.99
SC-MG-G10   273     SIM Cubicle         16.99
DB-SG-G01   266     Mediocre Kingdoms   19.99
```

You can also rename the column name when it is added to the events. See the following example:

```
index=main sourcetype=access_combined_wcookie action=purchase
| top 3 productId showperc=f
| lookup prices.csv productId AS productId OUTPUT product_name, sale_price AS final_price
```

CHAPTER 6   USING LOOKUPS

Splunk renames the column name *sale_price* as *final_price* when it adds it to the events. Note that it does not change the lookup table contents. It just changes the events displayed:

```
productId       count       final_price     product_name
-----------------------------------------------------------
WC-SH-G04       275         19.99           World of Cheese
SC-MG-G10       273         16.99           SIM Cubicle
DB-SG-G01       266         19.99           Mediocre Kingdoms
```

As you can see, the *sale_price* has been renamed to final_price when the events are displayed. Next, we'll take a look at how to maintain the contents of the lookup table.

Maintaining the Lookup

A common use case of lookups is enriching your events with external data that are static in nature. For example, HTTP status code descriptions can be used to add additional information to your events based on the HTTP status code. Consider the following example:

```
index=main sourcetype=access_combined_wcookie status = 5*
| stats count by status
| lookup apache_httpstatus_lookup status AS status OUTPUT status_
description
```

The SPL query uses the lookup table *apache_httpstatus_lookup* provided by Apache TA (Technical Add-on) which can be downloaded from https://splunkbase.splunk.com. It looks up status code and pulls the status description. The output looks like the following:

```
status      count       status_description
---------------------------------------------------
500         733         Internal Server Error
503         952         Service Unavailable
505         480         HTTP Version Not Supported
```

The output is very informative as it shows what a particular status code means. The status code and status description are static in nature, and they seldom change. But at times you may run into situations where you want to keep your lookup tables updated periodically. For example, in the *prices.csv* file, we may need to update the sale price of a product if it goes on sale. One way to update the lookup table is to simply replace the file by uploading a new file. While this will work, it can be a lot of manual work if the changes are frequent.

Splunk provides the outputlookup command which can be used to create or update a lookup table using the results of a search. Let's take a look at this command in detail.

Using the outputlookup Command

Using the outputlookup command, you can use the results of a search to create or update a lookup table. This can be very handy if your lookup table needs to be updated periodically. The basic syntax of the outputlookup command is as follows:

```
...| outputlookup override_if_empty=<t|f> <lookup-table-name>
```

The command accepts the lookup table name as a parameter. When there are no search results to be piped to the outputlookup command, the target lookup table is deleted if it already exists. This can be dangerous as you may accidently delete your lookup table. I recommend using the override_if_empty=f option to prevent this. Let us see an example:

To create a new lookup table with five worst-performing vendors and their number of sales:

```
index=main sourcetype=vendor_sales
| rare 5 VendorID showperc=f countfield="Number_of_Sales"
| outputlookup override_if_empty=f  low_performing_vendors.csv
```

The SPL query retrieves the five worst-performing vendors by using the rare command and changes the count field name to *Number_of_Sales* to be descriptive. The output of the rare command is sent to the outputlookup command which creates a new lookup table named *low_performing_vendors.csv*. You can verify this by using inputlookup command:

```
| inputlookup low_performing_vendors.csv
```

Splunk produces the following output:

```
Number_of_Sales    VendorID
-----------------------------
16                 9115
17                 1270
19                 1287
19                 5062
20                 1224
```

From this point, you can use the lookup table low_performing_vendors.csv in your other searches using the lookup command. Now, what about updating an existing lookup table? My advice for using `outputlookup` command to update an existing table is to not use it for simply updating a portion of the table. You need to be able to replace the **entire content of the lookup table**. This is because by default, the columns that are not in the results are removed from the lookup table. This can result in surprising truncation of your lookup table. In the next section, let's review some of the best practices when utilizing lookups.

Lookups Best Practices

While utilizing lookups, there are a few best practices you should follow to maximize the use:

1. Pay attention to the size of your lookup tables. Larger lookup tables adversely affect the search performance in distributed search setups. Work with your Splunk administrator to ensure your lookup tables are not going to be detrimental to the Splunk platform.

2. When using `outputlookup` command to update a lookup table, the entire lookup table must be replaced.

3. Create lookup tables in such a way that the key field (match field) is the first (leftmost) field to improve performance.

In the next section, let's take a look at how to make lookups automatic.

CHAPTER 6 USING LOOKUPS

Creating Automatic Lookups

You can make the lookup automatic for commonly used fields. In this way, you can avoid invoking the lookup command manually. For example, in order to populate product name and sale price, you can create an automatic lookup using the *prices.csv* file we uploaded earlier. You can create an automatic lookup in Splunk Web by navigating to *Settings* ➤ *Lookups* ➤ *Automatic Lookups*. Click *New Automatic Lookup*. As shown in Figure 6-6, enter the values.

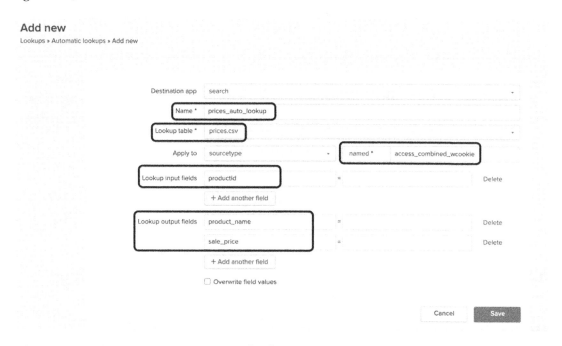

Figure 6-6. *Creating an automatic lookup*

For the *Name* field, you can provide any meaningful name. The lookup table must be selected from the drop-down. In the *Apply to* field, choose *sourcetype* and enter *access_combined_wcookie*. Then, you have to specify the lookup input field that you want to match. In our case, it is the *product Id*. Next, you will have to provide the output field you would like to retrieve. Enter *product_name* and *sale_price*. Click Save.

In order to invoke the automatic lookup, all you have to do is to use the output fields in your SPL. For example, consider the following SPL:

```
index=main sourcetype=access_combined_wcookie action=purchase
| stats values(product_name) AS product_name , sum(sale_price) AS total_sales by productId
```

Notice the use of *product_name* and *sale_price* fields. The preceding SPL calculates the sum of sale prices for products purchased. An excerpt of the output is shown in the following:

```
productId        product_name         total_sales
---------------------------------------------
BS-AG-G09        Benign Space Debris  3018.49
CU-PG-G06        Curling 2014         2514.52
DB-SG-G01        Mediocre Kingdoms    5317.34
DC-SG-G02        Dream Crusher        5647.74
```

Caution Because automatic lookups run behind the scenes, in large Splunk environments with hundreds of users, automatic lookups created by one user may affect the search results of the other (when the knowledge objects are shared). So, be cautious while using automatic lookups.

That brings us to the conclusion of this chapter on lookups. Let's review the key takeaways from this chapter.

Key Takeaways

Lookup is a key component of Splunk that enables you to enrich your search results with external data. Learning to correctly create and use lookups will help in adding useful information to your Splunk reports and dashboards. CSV files are the most common type of lookups. Here are the key takeaways from this chapter:

1. A lookup enables you to add data to your search results that can help in adding more meaning to your reports.

2. You can upload a CSV file to be used as a lookup.

3. The CSV file must have a header line. The header columns are used as fields in Splunk.

4. You must have one or more fields from your events that match the columns in the lookup table.

5. You can use `inputlookup` command to verify the contents of the lookup.

6. The `lookup` command is used to perform the lookup. You need at least one field to match the events with the lookup.

7. By default, for a given match, all fields from the lookup except the match field are added to the event.

8. You can use `outputlookup` command to create or update lookup tables based on search results.

9. Use `override_if_empty=f` if you don't want your lookup to be deleted if the search result is empty.

10. Pay attention to the size of your lookup tables, and work with your Splunk administrator for optimizing.

Now that you have learned about lookups and their usage, in the next chapter, we'll take a look at some of the advanced SPL commands.

CHAPTER 7

Advanced SPL Commands

What you have learned so far about SPL is more than enough to make you look like a Splunk ninja. But there are many commands in SPL that may require a steeper learning curve. They solve varieties of problems, so it is worth learning them. For example, when a field has multiple values, it is treated as a multivalued field. There are host of commands to effectively make use of these multivalued fields. And there are commands that use machine learning algorithms to predict future values of a field and uncover text patterns from your data. In most cases, by learning the basic syntax and few options of these commands, you should be able to make use of them immediately. In this chapter, we'll look at many of these commands with detailed examples. By the end of this chapter, you will have learned enough about these commands that you should be able to start using them right away.

predict

You can use the `predict` command to forecast the future value of one or more numeric fields based on the past data. The command must be preceded by the `timechart` command. For example, to predict the number of 500 series HTTP errors (internal server errors) based on the access log data

```
index=main sourcetype=access_combined_wcookie status=5*
| timechart span=1d count AS "Server_Errors"
| predict "Server_Errors" AS Predicted_Server_Errors
```

The SPL query retrieves the 500 series HTTP errors and pipes the results to `timechart` command to plot the count over time. It then pipes the results to the `predict` command with the field to perform the prediction (*Server_Errors*). The result produces three new fields – *Predicted_Server_Errors, lower95(Predicted_Server_Errors),* and *upper95(Predicted_Server_Errors).* See Figure 7-1.

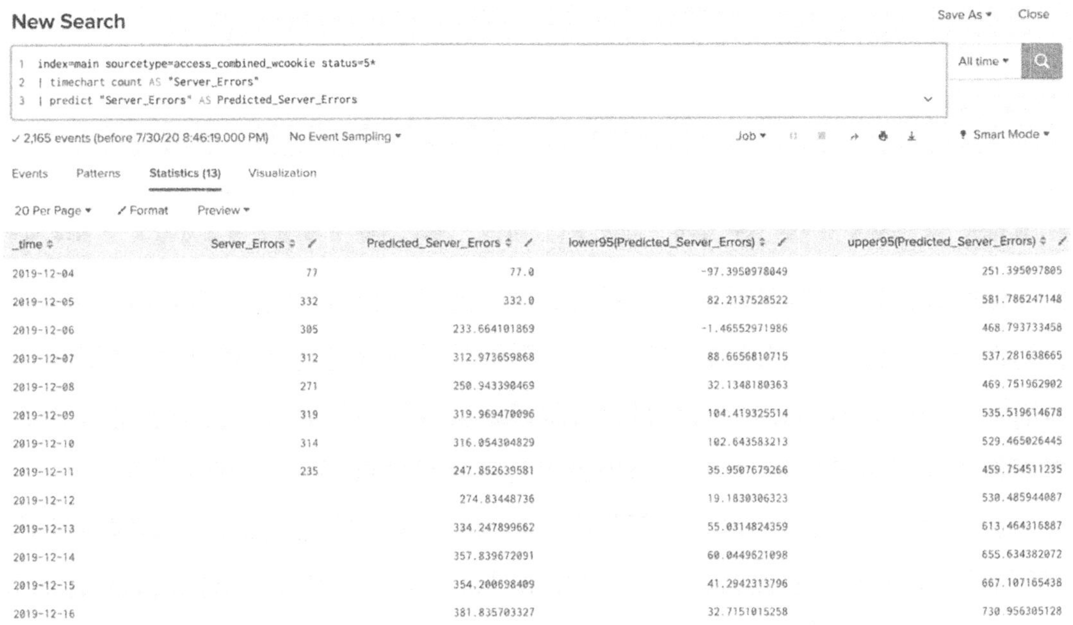

CHAPTER 7   ADVANCED SPL COMMANDS

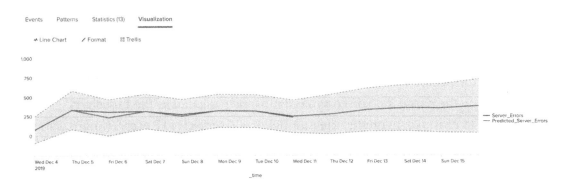

Figure 7-1. Using the predict command

The visualization produced by the preceding result is shown in Figure 7-2.

Figure 7-2. Visualization produced by the predict command

The predict command estimates the future value of *Server_Errors* with 95% confidence interval. The upper and lower 95th percentile values are calculated for each data point. It supports several algorithms for computing the estimated values. By default, it predicts five future time spans. You can override this by using *future_timespan* option. For example:

CHAPTER 7 ADVANCED SPL COMMANDS

```
index=main sourcetype=access_combined_wcookie status=5*
| timechart span=1d count AS "Server_Errors"
| predict "Server_Errors" AS Predicted_Server_Errors future_timespan=10
```

The SPL query will forecast the data for the next ten days (since one time span is equal to one day in the query) following the last known data point. See Figure 7-3.

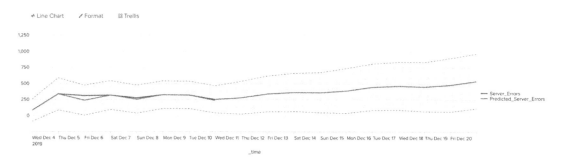

Figure 7-3. *Extending the future time span with predict command*

Note The result of predict command is probabilistic expectation, and the actual results may not match the expectation exactly.

Next, let's take a look at kmeans command.

kmeans

kmeans is another advanced command that you can use in numerical data analysis. Specifically, kmeans divides the numerical data into a number of clusters. The number of clusters is denoted by k, and you can specify the value of k as a parameter in the command. For each cluster, kmeans calculates a centroid value of the specified fields. Then, it groups events closest to the centroid values into respective clusters. Events belonging to the same cluster are moved next to each other. The command creates a field named *CLUSTERNUM* that indicates the cluster. For example, using Splunk tutorial data, in order to create four clusters of products based on their sale price

```
index="main" sourcetype="access_combined_wcookie" action=purchase
| kmeans k=4 sale_price
| stats values(product_name),avg(sale_price) by CLUSTERNUM
```

Splunk produces the following result as shown in Figure 7-4.

CLUSTERNUM	values(product_name)	avg(sale_price)
1	Benign Space Debris Grand Theft Scooter Mediocre Kingdoms World of Cheese	20.154456233421563
2	Dream Crusher Manganiello Bros. Orvil the Wolverine Pony Run	25.01773246329508
3	Curling 2014 Final Sequel SIM Cubicle	16.989999999999824
4	Fire Resistance Suit of Provolone Holy Blade of Gouda Manganiello Bros. Tee Puppies vs. Zombies World of Cheese Tee	4.276604361370612

Figure 7-4. *Using kmeans to create clusters based on numerical data*

The SPL query creates four clusters based on the sale price of the product. You can change the number of clusters by using a different value for k. Each event in the result is added with a cluster number (*CLUSTERNUM* field) and the centroid value. The `stats` command lists the product names and average sale price ground by the cluster number. You can specify a custom name for the cluster number field (which is CLUSTERNUM by default) by using the *cnumfield* option. Next, let's take a look at another command that can help in grouping events.

cluster

The `cluster` command groups events based on patterns it detects in the text. This is unlike kmeans command which operates on numerical data. The cluster command breaks the data into terms and computes cosine similarity between the events. It adds two new fields to the result – *cluster_label* and *cluster_count*. The field names can be customized. It is generally useful to pipe the results of the cluster command to table command and print only the relevant fields. For example, to find the most common text patterns in the secure.log in Splunk tutorial data

```
index="main" sourcetype="secure"
| cluster t=0.5 showcount=t
| sort -cluster_count
| table cluster_label, cluster_count, _raw
```

CHAPTER 7 ADVANCED SPL COMMANDS

The SPL query applies the `cluster` command on the data retrieved from main index, limiting results to sourcetype *secure*. It uses two options with the command. The first option t specifies the sensitivity of the grouping. Its value must be greater than 0 and less than 1. Higher sensitivity (close to 1) results in more clusters; lower sensitivity (close to 0) results in fewer clusters. In other words, the higher the sensitivity, the more similar the events have to be for them to be considered in the same cluster. The next option *showcount=t* forces the cluster count to be added to each event. We need this to show the number of events in each cluster. We then sort the results based on the size of the cluster (number of events) and use the `table` command to print the results. Figure 7-5 shows the result.

cluster_label	cluster_count	_raw
1	33031	Wed Dec 10 2019 00:15:03 www3 sshd[3509]: Failed password for invalid user vpxuser from 49.212.64.138 port 4805 ssh2
2	3590	Wed Dec 10 2019 00:15:03 www3 sshd[45881]: pam_unix(sshd:session): session opened for user djohnson by (uid=0)
4	2365	Tue Dec 09 2019 00:15:03 www3 sshd[43861]: Server listening on :: port 22.
5	760	Thu Dec 11 2019 00:15:02 www1 sshd[55981]: Received disconnect from 10.2.10.163 11: disconnected by user
3	341	Wed Dec 10 2019 00:15:03 www3 sudo: djohnson ; TTY=pts/0 ; PWD=/home/djohnson ; USER=root ; COMMAND=/bin/su

Figure 7-5. *Using cluster command to find text patterns*

The `cluster` command can be very useful to find unnecessary noise in the events. For example, after a release of a new version of your application, you can run the `cluster` command against your log files to uncover any new recurring errors. Another point to note is that you can also focus on a particular field to cluster by using the *field* option. If you don't specify a field name, _raw is processed. For example:

...| cluster t=0.5 **field=product_name**

The *Patterns* tab in Splunk Web uses the cluster command to discover patterns in the search results. Figure 7-6 shows the Patterns tab.

CHAPTER 7   ADVANCED SPL COMMANDS

Figure 7-6. *Using cluster command to find text patterns*

Tip   Review the Patterns tab especially after a change in your application, for example, a release of a new version. If there are new, repeating error messages introduced as part of the release, Patterns tab will reveal them.

The Patterns tab has a few additional functionalities. In addition to discovering patterns, when you select a pattern, Splunk also shows the keywords that were used. You can use these keywords to search for these events. Further, you can save this event as event type or even create an alert right from the Patterns tab. There is a slider at top that you can use to change the sensitivity of pattern matching. Sliding toward the left creates more patterns and toward right creates fewer patterns. Next, let's take a look at how to remove *outliers* from your data.

Outlier

Outlier is a data point that differs significantly from other observations. Having outliers in your data can skew the results if they are not removed. For example, if you are analyzing market segmentation, not removing the outliers can lead to incorrect conclusions. Splunk provides the command `outlier` to remove outlying numerical values. You can specify which fields you want to process, or Splunk will try to process all fields.

CHAPTER 7   ADVANCED SPL COMMANDS

Outlying values are determined by using interquartile range (IQR) which is computed from the difference between 25th percentile and 75th percentile of the numerical value. You can define a parameter to control the threshold used for outlier detection. By default, this parameter, named *param,* is 2.5. If a given data point is less than *25th percentile – param*IQR,* or greater than *75th percentile + param*IQR,* it is considered as an outlier. You can either completely remove the outlier or transform it to be equal to the threshold. As shown in Figure 7-7, consider the `timechart` of average response time from a hypothetical access log:

```
...| timechart span=3m avg(response_time) AS avg_response_time
```

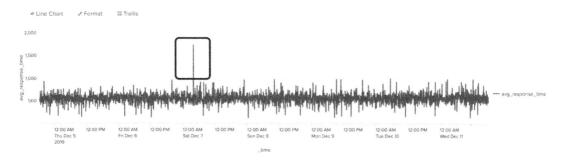

Figure 7-7. Outliers in time-series data

You can clearly see that there is an outlying spike. If you want to remove this outlying data point, all you have to do is pipe the result into the `outlier` command:

```
...| timechart span=3m avg(response_time) AS avg_response_time
| outlier avg_response_time action=remove
```

The resulting `timechart` is shown in Figure 7-8.

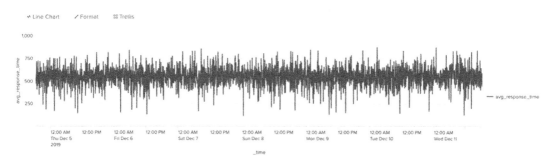

Figure 7-8. Time-series data after the removal of outliers

CHAPTER 7　ADVANCED SPL COMMANDS

As you can readily see, removing outliers provides a better picture of the time-series metrics. With outliers removed, you can see the upper- and lower-bound values more accurately.

The *action* option of the `outlier` command lets you either remove the value completely or transform it to be equal to the threshold value. The default action is to transform.

Next, let's take a look at `fillnull` and `filldown` commands that enable you to handle null values effectively.

fillnull and filldown

As you work with machine data, you will observe that a common issue to deal with is null values. A field is considered null when the value for the field is completely missing. There are many reasons a field can be set to null. For example, your raw data may not have all the values that you need in the event, or your field extractions may fail due to a change in the raw data, and so on. Whatever may be the cause, you may want to correctly handle the null values in order to produce accurate calculations.

By default, the command `fillnull` will replace all the null values with 0. You can override this by using the *value* option. You can specify a field list to process. If you don't specify a field list, `fillnull` command will process all fields. For example, consider the access logs from the Splunk tutorial data. When the *action* field does not have any value (i.e., null), you can replace the null value with "N/A" using the following SPL query:

```
index=main sourcetype=access_combined_wcookie
| fillnull action value="N/A"
| stats count by action
```

Splunk produces the following result:

```
action              count
-------------------------
N/A                 19814
addtocart            5743
changequantity       1402
purchase             5737
remove               1445
view                 5391
```

If you want to replace all null values with 10, you can use the following SPL query:

```
...| fillnull value=10
```

Instead of filling the null value with 0 or your own replacement, you can have Splunk use the last nonnull value for a field. If there is no previous nonnull value for a field, the field is left blank. Consider the following hypothetical data:

```
server      response_time
-----------------------
pluto       345
earth       231
earth
venus
pluto       65
earth       438
saturn
```

If you want to fill in the null values for the *response_time* field with the previous nonnull value, you can use the following SPL query:

```
...| filldown response_time
```

Splunk will produce the following:

```
server      response_time
-----------------------
pluto       345
earth       231
earth       231
venus       231
pluto       65
earth       438
saturn      438
```

Notice how the null values were replaced by the previous nonnull values. Next, let's look at a command that is very useful in converting text - the convert command.

convert

You can use the `convert` command to transform varieties of text data into numerical or human-readable format. You can use the AS clause to create new fields to store the new values. If you don't use the AS clause, original values are replaced. The command expects at least one conversion function as a parameter.

To convert Unix time to human-readable time, you can use the *ctime* function of convert command. For example, to convert the value of *_indextime* field, which stores the time at which an event was indexed, into human-readable form

```
index=_internal log_level=ERROR
| convert ctime(_indextime) AS IndexTime
| table IndexTime, _time
| head 3
```

The SPL query retrieves the ERROR messages from *_internal* index, converts the *_indextime* into human-readable form, and stores the result in `IndexTime` field. The `head` command limits the number of results returned to 3. The result is shown in the following:

```
IndexTime                _time
------------------------------------------
08/01/2020 11:10:58      2020-08-01 11:10:57.472
07/30/2020 22:02:03      2020-07-30 22:02:02.274
07/30/2020 21:35:46      2020-07-30 21:35:45.528
```

Notice the format of the timestamp in the *IndexTime* column. You can use the *timeformat* option to change the format to match *_time*:

```
index=_internal log_level=ERROR
| convert timeformat="%Y-%m-%d %H:%M:%S.%3N" ctime(_indextime) AS IndexTime
| table IndexTime, _time
```

Splunk produces the following result:

```
IndexTime                    _time
----------------------------------------------------
2020-08-01 11:10:58.000      2020-08-01 11:10:57.472
2020-07-30 22:02:03.000      2020-07-30 22:02:02.274
2020-07-30 21:35:46.000      2020-07-30 21:35:45.528
```

CHAPTER 7 ADVANCED SPL COMMANDS

To convert a human-readable time to Unix time, you can use the *mktime* function:

```
| makeresults
| convert mktime(_time) AS TimeInEPOCH
```

Notice the use of `makeresults` command which simply returns one event with the timestamp of when the command was run. The `convert` command converts the human-readable _time to Unix time and stores in the new field *TimeInEPOCH*. The result is shown in the following:

```
TimeInEPOCH            _time
-----------------------------------
1596316896             2020-08-01 16:21:36
```

Another useful function is `dur2sec`, which calculates the number of seconds from time in the format HH:MM:SS. For example:

```
| makeresults
| eval duration="4:34:05"
| convert dur2sec(duration) AS durationInSeconds
| table duration,durationInSeconds
```

The SPL query creates a new field named *duration* and stores *4:34:05* in it. The convert function calculates the equivalent seconds and stores it in *durationInSeconds* field. The results look like the following:

```
duration    durationInSeconds
-----------------------------
4:34:05     16445
```

When you are working with financial data, often you come across numbers that are separated by commas, such as 23,456. While it makes the number easily readable, for mathematical calculations, you may want the value to be represented as a number, with commas removed. The function *rmcomma* can be used for this purpose:

```
| makeresults
| eval total_sales = "2,345,895.89"
| convert rmcomma(total_sales) AS total_sales_num
| table total_sales,total_sales_num
```

203

CHAPTER 7   ADVANCED SPL COMMANDS

The SQL query removes the commas from *total_sales* and stores the result in the *total_sales_num* field. The result is shown in the following:

```
total_sales      total_sales_num
-------------------------------
2,345,895.89     2345895.89
```

> **Note**   You can achieve the reverse, converting a number to string using eval *toString* function, ...| eval total_sales_str = toString(total_ sales,"commas").

Next, we'll learn about handling multi valued fields.

Handling Multivalued Fields

Until now we've been seeing fields with single values, for example, *action=purchase*, *status=200*, and so on. But there are situations where a field can have multiple values. Email recipient list is a classic example of a multivalued field. A multivalued field has more than one value. The SPL commands that work with multivalued fields are makemv, mvcombine, mvexpand, and nomv. Further, there are eval functions that help with multivalued fields, *mvcount()*, *mvfilter()*, *mvindex()*, *mvsort()*, *mvzip()*, and *mvjoin()*.

makemv

The command makemv converts a single-valued field into a multivalued field. Consider the following events:

```
[11/Dec/2019:18:18:58] sending email to jbarnes@acmecorp.com,ksmith21@
acmecorp.com,rmanroe@acmecorp.com
[11/Dec/2019:19:06:13] sending email to sjackson4@acmecorp.com,hbernstien@
acmecorp.com
[11/Dec/2019:19:18:23] sending email to cthomson@acmecorp.com
```

You can extract the email addresses in a field using the `rex` command as shown in the following SPL query:

```
...| rex "sending email to (?<Email_To>[^$]+)"
| table _time,Email_To
```

The SPL uses `rex` command to capture everything after the string *sending email to* until the end of line. The character class [^$]+ refers to any character that is not the end of the line. The resulting looks like the following:

_time	Email_To
2019-12-11 18:18:58	jbarnes@acmecorp.com, ksmith21@acmecorp.com, rmanroe@acmecorp.com
2019-12-11 19:06:13	sjackson4@acmecorp.com, hbernstien@acmecorp.com
2019-12-11 19:18:23	cthomson@acmecorp.com

But the problem here is, as you can see, the *Email_To* field does not handle the individual email addresses. Instead, it is simply a sequence of strings. For example, if you need to find out the total number of unique email addresses, you may try to use the following SPL query:

```
...| rex "sending email to (?<Email_To>[^$]+)"
| stats dc(Email_To) AS Distinct_Email_Addresses
```

The `stats` function *dc* stands for distinct count, and it returns the unique value of the specified field. But the answer you get won't be correct. The SPL query outputs the following result:

```
Distinct_Email_Addresses
-----------------------
3
```

As you can see, instead of 6 distinct email addresses, the result we got is 3. In order to handle the email addresses individually, we need to handle the *Email_To* field as a multivalued field. We can use the `makemv` command to achieve this:

```
...| rex "sending email to (?<Email_To>[^$]+)"
| makemv delim="," Email_To
| table _time,Email_To
```

The SPL query uses the `makemv` command to split the value of *Email_To* into multiple values using the delimiter ",". The resulting values of *Email_To* field will look like the following:

```
_time                    Email_To
-------------------------------------------------
2019-12-11 18:18:58      jbarnes@acmecorp.com
                         ksmith21@acmecorp.com
                         rmanroe@acmecorp.com
2019-12-11 19:06:13      sjackson4@acmecorp.com
                         hbernstien@acmecorp.com
2019-12-11 19:18:23      cthomson@acmecorp.com
```

Also, if you run the query to find the distinct email addresses, you will get the correct answer:

```
| rex "sending email to (?<Email_To>[^$]+)"
| makemv delim="," Email_To
| stats dc(Email_To) AS Distinct_Email_Addresses
```

Splunk produces the following result:

```
Distinct_Email_Addresses
------------------------
6
```

Instead of using delimiter, you can also use a regular expression to split the values. If you need to use regular expression, the syntax is as follows:

```
...| makemv tokenizer=<regular expression>
```

CHAPTER 7 ADVANCED SPL COMMANDS

The regular expression must contain a capturing group that represents each value of the multi valued field. For example, if we have to use regular expression for our *Email_To* example

```
...| makemv tokenizer="([^,]+),?" Email_To
```

The regular expression contains a capturing group ([^,]+) that defines the value as any character that is not a comma. The following ,? indicates zero or more occurrence of comma. This regular expression is used repeatedly until all values are found.

nomv

The command nomv performs the reverse of makemv. It converts a multivalued field into a single-valued field. For example, the values function of stats command usually produces a multivalued field. Consider the following SPL query:

```
index=main sourcetype=access_combined_wcookie action=purchase
| stats values(productId) AS products by categoryId
| sort 2 categoryId
```

The SPL query finds the list of products grouped by *categoryId* for all purchases. It uses sort to list the first two categories. Figure 7-9 shows the result.

categoryId	products
ACCESSORIES	WC-SH-A01
	WC-SH-A02
ARCADE	BS-AG-G09
	FI-AG-G08
	MB-AG-G07

Figure 7-9. *Result of stats(values) which proudces a multivalued field*

The field *product* in the result is a multivalued field. If you use the nomv command on the *products* field, it is converted into a single-valued field:

```
index=main sourcetype=access_combined_wcookie action=purchase
| stats values(productId) AS products by categoryId
| nomv products
| sort 2 categoryId
```

CHAPTER 7   ADVANCED SPL COMMANDS

Figure 7-10 shows the result.

Figure 7-10. Converting a multivalued field into a single-valued field

The products field is no longer a multivalued field.

mvexpand

You can use the mvexpand command to create an event for each value in the multivalued field. For example, in our *Email_To* example, you can create a result for each email address using the following SPL query:

```
...| makemv delim="," Email_To
| mvexpand Email_To
```

Instead of showing three events, Splunk shows six events as shown in Figure 7-11.

Figure 7-11. Using mvexpand to create a result for each value of a multivalued field

mvcombine

The command mvcombine does the opposite of mvexpand. It combines events that are identical except for the specified field that contains a single value. It creates a single event with the specified field as a multivalued field. You can specify a delimiter that will be used to separate the values within the multivalued field. A basic example of mvcombine is shown in the following:

```
...| mvcombine delim="," host
```

The SPL query uses the mvcombine command to operate on events that have identical data except the *host* field. It combines the events into a single event with the host field formed as a multivalued field. As an example, consider the following table:

```
productId     categoryId
--------------------
DC-SG-G02     STRATEGY
DB-SG-G01     STRATEGY
PZ-SG-G05     STRATEGY
FS-SG-G03     STRATEGY
```

The *categoryId* column is exactly the same on each event. We can combine the four rows into one row using the mvcombine command as follows:

...| mvcombine delim="," productId

Splunk produces the following result:

```
productId     categoryId
--------------------
DB-SG-G01     STRATEGY
DC-SG-G02
FS-SG-G03
PZ-SG-G05
```

You may be wondering what happened to the delimiter ",". By default, the `mvcombine` command creates a multivalued version of the field (in our example, *productId*) and a single-valued version of the field. The multivalued version is displayed by default. You have to pipe the output to `nomv` to display the single-valued version as shown in the following:

```
...| mvcombine delim="," productId
| nomv productId
```

Output looks like the following:

```
productId                                categoryId
--------------------------------------------------
DB-SG-G01,DC-SG-G02,FS-SG-G03,PZ-SG-G05   STRATEGY
```

In addition to the commands, SPL provides a few useful eval functions to handle multivalued fields. Let's take a look at some of them next.

mvcount

The eval function `mvcount` returns the number of values in a multivalued field. For instance, in our *Email_To* example earlier, you can use the `mvcount` function to display the number of email addresses in each multivalued field:

```
...| makemv delim="," Email_To
| eval count=mvcount(Email_To)
| table _time,Email_To,count
```

Splunk produces the following results:

```
_time                   Email_To                    count
---------------------------------------------------------
2019-12-11 18:18:58     jbarnes@acmecorp.com          3
                        ksmith21@acmecorp.com
                        rmanroe@acmecorp.com
2019-12-11 19:06:13     sjackson4@acmecorp.com        2
                        hbernstien@acmecorp.com
2019-12-11 19:18:23     cthomson@acmecorp.com         1
```

mvindex

You can use the eval function `mvindex` to access a particular value in a multivalued field. The index numbers start from 0. The first value is at index 0, second at 1, and so on. Multivalued fields maintain the order of the values. For example, to access the last email address in the multivalued *Email_To* field

```
...| makemv delim="," Email_To
| eval count=mvcount(Email_To)
| eval last_email = mvindex(Email_To,count - 1)
| table _time,last_email
```

In order to access the last element in the multivalued field, we use *count – 1* as the index, as the index starts from 0 and goes up to *count – 1*. Splunk produces the following result:

```
_time                    last_email
-----------------------------------------
2019-12-11 18:18:58      rmanroe@acmecorp.com
2019-12-11 19:06:13      hbernstien@acmecorp.com
2019-12-11 19:18:23      cthomson@acmecorp.com
```

mvfilter

You can use the eval function `mvfilter` to filter a multivalued field using an arbitrary Boolean expression. For example, in our *Email_To* multivalued field, to keep the email addresses that end with son in the user ID portion

```
...| eval Emails_ending_in_son = mvfilter(match(Email_To,".*son.*"))
| table Email_To,Emails_ending_in_son
| fillnull value="No matching emails"
```

The SPL query uses `mvfilter` function with another eval function `match` to filter the email addresses. Note that `match` function accepts regular expressions. And the query also uses the `fillnull` command with a custom replacement value for null. The output looks like the following:

```
Email_To                      Emails_ending_in_son
-----------------------------------------------------
jbarnes@acmecorp.com          No matching emails
ksmith21@acmecorp.com
rmanroe@acmecorp.com
sjackson4@acmecorp.com        sjackson4@acmecorp.com
hbernstien@acmecorp.com
cthomson@acmecorp.com         cthomson@acmecorp.com
```

mvfind

The eval function `mvfind` searches for a value based on regular expression you provide and, if a match is found, returns the index of the first matched value. For instance, in our *Email_To* example, to find the index of the email address that has the string *smith*

```
...| eval smith_index = mvfind(Email_To,".*smith.*")
```

The SPL query will return 1 as ksmith21@acmecorp.com is the second entry in the first event in our example. If an event does not have a match, no value is returned to the *smith_index* field. Figure 7-12 shows the result.

Figure 7-12. *Using mvfind to find the index of a value in a multivalued field*

CHAPTER 7   ADVANCED SPL COMMANDS

mvjoin

You can use mvjoin to combine the values in a multivalued field with a specified delimiter. The following example joins all the individual email addresses using delimiter ";" (semicolon):

```
... | eval joined_emails = mvjoin(Email_To,";")
| table Email_To,joined_emails
```

Splunk produces the result as shown in Figure 7-13.

Email_To	joined_emails
jbarnes@acmecorp.com ksmith21@acmecorp.com rmanroe@acmecorp.com	jbarnes@acmecorp.com;ksmith21@acmecorp.com;rmanroe@acmecorp.com
sjackson4@acmecorp.com hbernstien@acmecorp.com	sjackson4@acmecorp.com;hbernstien@acmecorp.com
cthomson@acmecorp.com	cthomson@acmecorp.com

Figure 7-13. Using mvjoin to combine values in a multivalued field

mvsort

The mvsort() eval function sorts the values in a multivalued field. For example, to sort the email addresses in *Email_To* field

```
...| eval sorted_emails = mvsort(Email_To)
| table Email_To,sorted_emails
```

Splunk produces the result as shown in Figure 7-14.

Email_To	sorted_emails
jbarnes@acmecorp.com ksmith21@acmecorp.com rmanroe@acmecorp.com	jbarnes@acmecorp.com ksmith21@acmecorp.com rmanroe@acmecorp.com
sjackson4@acmecorp.com hbernstien@acmecorp.com	hbernstien@acmecorp.com sjackson4@acmecorp.com
cthomson@acmecorp.com	cthomson@acmecorp.com

Figure 7-14. Using mvsort to sort values in a multivalued field

split

The `split` function of `eval` creates a multivalued field by splitting a single-valued field with a delimiter you specify. For example:

```
| makeresults
| eval phone_number = "634-2389858"
| eval area_code = mvindex(split(phone_number,"-"),0)
| table phone_number,area_code
```

The SPL query creates a multivalued field using the `split` function to split the phone number using the delimiter "-". The `mvindex` function retrieves the value with the index number 0, which is the first value in the multivalued field. Splunk produces the following output:

```
phone_number      area_code
-------------------------
634-2389858       634
```

Next, let's take a look at how to handle structured data like JSON and XML during searching.

Extracting Fields from Structured Data

By default, Splunk will automatically extract key-value pairs from the raw data when the key-value pair is separated by equal sign "=", for example, *status=500*. In addition, if the data is of JSON format, Splunk will automatically extract the fields. If the data is in XML format, you can configure Splunk to automatically extract XML fields using *props.conf* (by setting KV_MODE=XML).

Note Splunk end users may not have access to *props.conf*. Splunk administrators usually own and manage *props.conf*.

CHAPTER 7 ADVANCED SPL COMMANDS

Let us take a look at how to make use of the automatically extracted json fields. Consider the following JSON data:

```
{
    "timestamp": 1596316896,
    "system": "web",
    "subsystem": "auth",
    "message": {
        "severity": "INFO",
        "id": 1234,
        "body": "Auth system initialized"
    },
    "email": [
        "ksmith21@acmecorp.com",
        "jbarnes@acmecorp.com"
        ]
}
```

When retrieving this data, it will be rendered by Splunk as shown in Figure 7-15.

Figure 7-15. *Rendering of json data in Splunk Web*

215

As you can see, Splunk automatically displays the data with syntax highlighting. In addition, it also extracts the fields and shows them in the INTERESTING FIELDS section on the left side. If you don't see the data correctly rendered, ensure that the data is indeed a complete JSON. You can use sites like jsonlint.com to validate JSON data. In addition, ensure your *search mode* is smart (or verbose). In fast mode, automatic field discovery is disabled.

You can access the JSON fields just like any other fields. For example, to retrieve events with INFO severity

```
index="main" sourcetype="sample_json" message.severity="INFO"
```

Note that the JSON field names follow the structure and hierarchy of the data. Arrays in JSON are denoted by a pair of curly braces, as in the case with *email{}* field. JSON arrays are stored as multivalued fields in Splunk.

Let's take a look at the `spath` command that helps with handling fields in structured data.

spath

The command `spath` enables to extract fields from structured data like JSON and XML. Consider the following sample of XML data:

```
<log>
<severity>INFO</severity>
<id>34532</id>
<timestamp>1596316896</timestamp>
<body>System startup complete</body>
<email><address>"ksmith21@acmecorp.com"</address><address>"jbarnes@
acmecorp.com"</address></email>
</log>
```

When you retrieve this event in Splunk, by default you will not see the XML elements extracted as fields. In order to extract fields from the XML elements, simply pipe the results to `spath` command as follows:

```
index="main" sourcetype="xmltest"
| spath
```

CHAPTER 7 ADVANCED SPL COMMANDS

As shown in Figure 7-16, Splunk extracts the XML elements as fields.

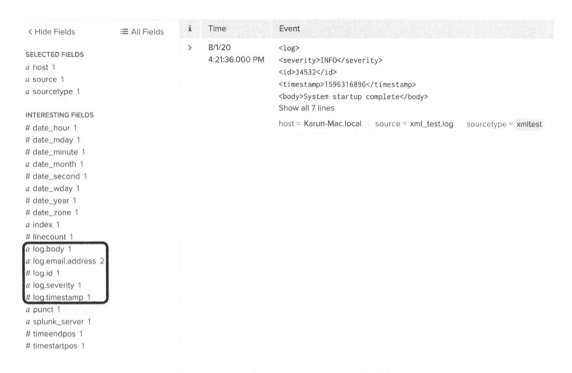

Figure 7-16. Using spath command to extract XML fields

The basic syntax of spath command is as follows:

spath input=<input field> output=<output field> path=<data path>

Input field defaults to *_raw* if one is not specified. Output field denotes the field to which the fields are extracted. If it is not specified, it defaults to the path of the data. Data path refers to the path of the data you want to extract. If it is not specified, all fields from the first 5000 characters of the input data are used for extraction.

Location of the data path is specified as a sequence of fields separated by period, for example, *log.severity*. The fields can also have optional index in curly brackets. If the index is an integer, it denotes the position of the data in an array. If it is a string preceded by an @ symbol, it denotes an XML attribute. For example, to retrieve the first email address from the sample data and store it in *email-1* field

```
index="main" sourcetype="sample_json"
| spath output=email-1 path="email{0}"
| table email-1
```

217

CHAPTER 7   ADVANCED SPL COMMANDS

The SPL query uses the `spath` command to retrieve the first element in email array and stores the result in *email-1* field. The result is shown in the following:

```
email-1
--------------------
ksmith21@acmecorp.com
```

Note that the index numbers start with 0 in JSON and 1 in XML data. Instead of specifying the index number, you can also refer to all the array elements by not specifying any number within the curly brackets. For example, consider the following JSON data:

```
{
   "bytes_read": 225,
   "type": "ad-hoc",
   "status": "success",
   "sourcetypes": {
       "xmltest": 1
   },
   "search_commands": [{
       "name": "timeliner",
       "duration": 0.002787516,
       "telemetryID": "14972441619928348640.0.0"
   }, {
       "name": "tags",
       "duration": 0.00000958,
       "telemetryID": "15849410590227250372.0.1"
   }]
}
```

To retrieve the names from all search commands in this sample data

```
| spath output=commands search_commands{}.name
| table commands
```

Splunk will produce the following result:

```
commands
-----------
timeliner
tags
```

Let's take a look at how array indexes work in XML data. Consider the following piece of XML:

```
<log>
<severity>INFO</severity>
<id>34532</id>
<timestamp>1596316896</timestamp>
<body>System startup complete</body>
<email><address inactive="no">"ksmith21@acmecorp.com"</address><address in
active="yes">"jbarnes@acmecorp.com"</address></email>
</log>
```

To retrieve the first email address and store it in a field named email-1

```
...| spath output=emails path=log.email.address{1}
```

To retrieve all the values of the attribute *inactive*

```
...| spath output=attributes path=log.email.address{@inactive}
```

To retrieve the attribute inactive specifically from the second email address

```
...| spath output=attributes path=log.email.address{2}{@inactive}
```

As a final note, sometimes instead of the entire event being JSON or XML, the data may be buried inside an event. In those cases, use the input option to specify the field name that contains the JSON or XML data. For example:

```
...| spath input=<field with json/xml data>
```

This concludes our discussion on handling structured data in Splunk. Note that it is best to have Splunk automatically extract the fields wherever possible. In order to achieve that, your data must be valid JSON/XML format. The command `spath` provides lots of flexibility in exploring and creating reports from structured data.

CHAPTER 7   ADVANCED SPL COMMANDS

Key Takeaways

In this chapter, we learned some of the advanced SPL commands that help you solve many real-world problems. The commands ranged from predictive analytics to handling multivalued fields to handling structured data. These commands can be at times confusing and hard to learn. But with practice, it will become easier and valuable in your Splunk journey. Here are the key takeaways from this chapter:

1. For forecasting a numerical value such as number of errors or average response times, use `predict` command. You can try using various prediction algorithms that the command supports.

2. In order to group the events into clusters based on their numerical values, use `kmeans` command. You can specify the number of clusters by using the k=<number of clusters> option.

3. Use the `cluster` command to identify text patterns in your data. Splunk Web's patterns tab internally uses the cluster command.

4. You can remove or transform outlying numerical data using `outlier` command.

5. When dealing with null values, you can use the `fillnull` command to customize the value to be replaced for null.

6. The command `filldown` lets you replace null with previous nonnull value.

7. The command `convert` transforms numerical data into human-readable text data.

8. You can use the `mvexpand` command to create individual events for each value in a multivalued field.

9. Eval provides many multivalued functions such as `mvcount` and `mvfilter` to retrieve the number of values from a multivalued field.

10. The command `spath` lets you retrieve fields from structured data such as JSON or XML.

The SPL commands you have learned so far are very frequently used, almost on a daily basis. But there are a few commands in SPL that may not be frequently used. This is not to say that they are not useful. In fact, some of these less-frequently used commands can be very helpful in certain situations, for example, to plot metrics in world map using geographical data. In the next chapter, let me introduce you to some of these commands and how they can help with your Splunk journey.

CHAPTER 8

Less-Common Yet Impactful SPL Commands

There are more than 140 commands in SPL to help you with numerous use cases pertaining to your machine data. We've seen many of these commands already. Yet, we haven't covered the entire SPL repertoire. You do not need to master all the commands to get the best out of Splunk. In fact, the commands we've learned so far will be more than enough for solving most problems. There are a few commands in SPL that you may not use on your day-to-day activities. While these are generally less common, that is not to say that they are not useful. In fact, some of these commands can be just what you need for your specific use case. In this chapter, I'll introduce somewhat-less-common yet impactful SPL commands. For each command, I'll explain its functionality and proceed to explain with an example or two.

iplocation

The command `iplocation` can be used to retrieve geographic location data based on IP address found in your events. This command is very useful if you manage public facing websites, provided you are capturing the client IP in your machine data. Client IP data is most commonly found in the web access logs. `iplocation` can derive the geographic location, including city, region, and country and add them to your events.

The command uses a third-party location file that ships with Splunk to derive the location information. The location file is named *GeoLite2-City.mmdb* and is located in SPLUNK_HOME/share. The basic syntax of the command is as follows:

```
...| iplocation <ip address field>
```

CHAPTER 8 LESS-COMMON YET IMPACTFUL SPL COMMANDS

The ip address field is the name of the field that has the client ip address. For example, to retrieve geographic location of clients based on the Splunk tutorial data

index=main sourcetype=access_combined_wcookie
| **iplocation clientip**

The SPL query uses the `iplocation` command with *clientip* as the field containing the ip address of the remote client. The resulting events are added with city, country, lat, lon, region. Figure 8-1 shows the fields that are automatically added.

Figure 8-1. Using iplocation command

Note The third-party geolocation database (Maxmind) that ships with Splunk is updated only when Splunk software is upgraded. Organizations can choose to use the paid version of Maxmind database if more frequent updates are preferred.

To create a pie chart showing the breakdown of web traffic by country

index=main sourcetype=access_combined_wcookie
| iplocation clientip
| **stats count by Country**

224

CHAPTER 8 LESS-COMMON YET IMPACTFUL SPL COMMANDS

When you choose the pie chart visualization, Splunk produces the following pie chart.

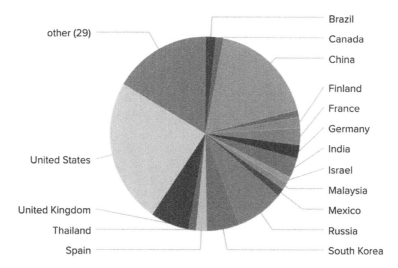

Figure 8-2. A pie chart that utilizes the iplocation command

Note With `allfields=true` option, the `iplocation` command also retrieves *Metrocode* and *Timezone* if available.

geostats

The command `geostats` calculates statistics and clusters them into geographical bins that can be plotted in a world map. This can be very useful if you want to generate a visualization to immediately understand the geographical distribution of a metric. In order for `geostats` to work, the data must have latitude and longitude information in the fields *lat* and *lon,* respectively. The `iplocation` command we looked at earlier produces these two fields based on ip address. It is common for `geostats` command to be preceded by `iplocation` command.

You can use statistical aggregation functions similar to `stats` command. For example, to plot the geographical distribution of the web traffic using Splunk tutorial data

```
index=main sourcetype=access_combined_wcookie
| iplocation clientip
| geostats count by Country
```

225

CHAPTER 8   LESS-COMMON YET IMPACTFUL SPL COMMANDS

When you choose cluster map visualization, Splunk produces a map as shown in Figure 8-3.

Figure 8-3. *A cluster map plotted in a map using geostats command*

You can use the zooming control on the map to focus on a particular area of the map. The map automatically adjusts to the location zoomed and displays granular statistics.

untable

The command untable takes tabular data and converts them to stats-style output. The basic syntax of unable is as follows:

...| untable <x-field> <y-field> <data field>

x-field is the field to be used as x axis; y-field is the field to be used as labels for the data series. The data field has the data to be converted. Consider the following data table:

```
action      ARCADE     SPORTS
--------------------------
purchase    537        148
view        220        65
```

The table lists the number of purchases and views for game genres ARCADE and SPORTS. This is similar to a `chart` output. In order to convert this into stats-style output, which produces one row for each combination of action and genre, you can use the following SPL:

```
...| untable action,category,count
```

Spunk produces the following output:

```
action     category    count
---------------------------
purchase   ARCADE      537
purchase   SPORTS      148
view       ARCADE      220
view       SPORTS      65
```

Output like this may be useful and more intuitive to understand than the chart-style output.

xyseries

The command `xyseries` performs the opposite of `untable` command. It reformats stats-style data into chartable tabular data. In this respect, it works exactly like chart command. The basic syntax of the command is as follows:

```
...| xyseries <x-field> <y-field> <data field>
```

x-field is the field to be plotted in x axis; y-field is the field to be used as column labels. Data field is the field that contains the data to be charted. Consider the following data:

```
productId   itemId    count
---------------------------
WC-SH-A01   EST-11    76
WC-SH-A01   EST-16    81
WC-SH-A02   EST-11    96
WC-SH-A02   EST-16    111
```

The stats-style table shows the count of *productId* and *itemId* breakdown that were purchased. In order to show this in chart-style table, use the following SPL:

```
...| xyseries productId,itemId, count
```

Splunk produces the following output:

```
productId      EST-11     EST-16
---------------------------------
WC-SH-A01       76          81
WC-SH-A02       96         111
```

As you can observe, it is exactly like the output of chart command. In fact, you can use chart command in place of xyseries in most cases. In general, if you need to perform some postprocessing of data, use xyseries. Otherwise, use chart.

bin

The command bin puts numerical values into discrete sets or bins. You can specify the size of each bin using the span option. You can also specify the maximum number of bins using the bins option. If the span size you specify creates more bins than the maximum number of bins you specified, the bins option is ignored. For example, consider a past season's NBA scoring data set (obtained from www.kaggle.com/dansbecker/nba-shot-logs?select=shot_logs.csv). The following SPL uses bin:

```
...| stats sum(PTS) AS Points_Scored by player_name
| sort 10 -Points_Scored
| bin Points_Scored span=500
| stats list(player_name) AS Player by Points_Scored
```

The SPL command first obtains the top 10 players based on the total points scored and then divides the points into bins of size 500. It then lists the players belonging to each bin. The result looks like the following:

```
Points_Scored       Player
-----------------------------------
1000-1500           stephen curry
                    James harden
                    klay thompson
```

	lebron james
	mnta ellis
500-1000	kyrie irving
	damian lillard
	lamarcus aldridge
	nikola vucevic
	chris paul

This sort of grouping can readily reveal insights about player's grouping. As another example, using Splunk tutorial data

```
index="main" sourcetype="access_combined_wcookie" action=purchase
| stats count AS "Purchases" by categoryId
| bin Purchases span=300
| stats list(categoryId) AS Category by Purchases
```

The SPL command groups game categories by the number of purchases. It uses the bin size of 300. The output looks like the following:

Purchases	Category
0-300	NULL
	SHOOTER
	SIMULATION
	SPORTS
300-600	ACCESSORIES
	ARCADE
	TEE
600-900	STRATEGY

Again, this sort of grouping readily reveals the best-performing and worst-performing group of games. The bin command can also be used for bucketing time. This can be handy if you want to create statistics grouped by time. For example:

```
...| bin span=1d _time
| stats count by _time
```

> **Note** The command `bucket` is an alias to `bin` command.

tstats

The command `tstats` is one of the most powerful commands you will ever use in Splunk. It calculates statistics using TSIDX files, typically created by accelerated data modes and indexed fields. The syntax of `tstats` can be a bit confusing at first. But after seeing a few examples, you should be able to grasp the usage.

`tstats` is a generating command, meaning it has to be the first command in the search pipeline with a preceding pipe. You can use aggregation functions just like `stats`. For example, the following command simply returns the total number of entries in the TSIDX files:

```
| tstats count
```

Like stats, it produces one line of output when it is not split by the by clause:

```
count
-----
253755
```

Using where and by Clause

You can use where clause and by clause as shown in the following:

```
| tstats count where index=* by index
```

This SPL query retrieves the number of entries in all indexes and breaks it down by the indexes. The output looks like the following:

```
index         count
-----------------
main          4849585
web           3758583
app           8402342
test          2341
```

CHAPTER 8   LESS-COMMON YET IMPACTFUL SPL COMMANDS

Querying Against Accelerated Data Models

In addition to querying indexed fields in TSIDX files, `tstats` can also query accelerated data models (which are stored in its own TSIDX files). In order to query an accelerated data model, use the data model keyword:

```
| tstats count FROM datamodel=<datamodel name>
```

The datamodel name is the name of the accelerated data model. To access a field within the accelerated data model, you need to specify the full path of the field. The full path is of the form *datasetName.fieldName*. You can verify the full path either by looking at the definition of the datamodel or by using the `datamodel` command. For example, running the datamodel command against the datamodel *internal_audit_logs*

```
| datamodel internal_audit_logs
```

Locate the field you are interested in (by expanding *objects* ➤ *fields*), and the *owner* of the field as shown in Figure 8-4.

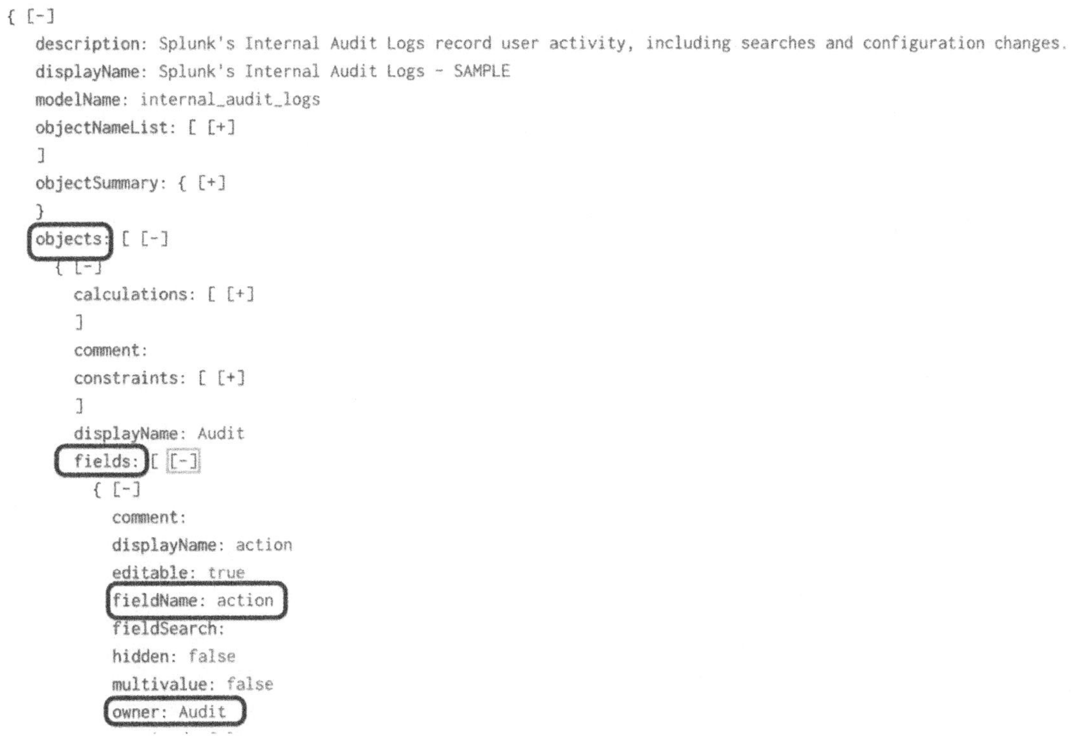

```
{ [-]
   description: Splunk's Internal Audit Logs record user activity, including searches and configuration changes.
   displayName: Splunk's Internal Audit Logs - SAMPLE
   modelName: internal_audit_logs
   objectNameList: [ [+]
   ]
   objectSummary: { [+]
   }
   objects: [ [-]
      { [-]
         calculations: [ [+]
         ]
         comment:
         constraints: [ [+]
         ]
         displayName: Audit
         fields: [ [-]
            { [-]
               comment:
               displayName: action
               editable: true
               fieldName: action
               fieldSearch:
               hidden: false
               multivalue: false
               owner: Audit
```

Figure 8-4. Using datamodel command to check the schema of a datamodel

CHAPTER 8   LESS-COMMON YET IMPACTFUL SPL COMMANDS

In this datamodel *internal_audit_logs*, which is a datamodel that comes with Splunk, the field *action* should be referred by the path *Audit.action*. See the following example on how this is used:

...| tstats count **FROM datamodel=internal_audit_logs BY Audit.action**

The SPL query uses `tstats` against the datamodel *internal_audit_logs* to pull the number of events grouped by the field *Audit.action*. An excerpt of the result is shown here:

```
Audit.action               count
-------------------------------
accelerate_search          3525
add                        3321
change_authentication      773
create_user                2
delete_by_keyword          5
```

You can use the option *summariesonly=true* to force `tstats` to pull data only from the *tsidx* files created by the acceleration. By default, if summaries don't exist, `tstats` will pull the information from original index. You will receive the performance gain only when `tstats` runs against the `tsidx` files. The following SPL shows how to force the *summariesonlhy* option:

| tstats **max(Audit.scan_count) summariesonly=t** FROM datamodel=internal_audit_logs **BY Audit.user**

In addition to setting *summariesonly=t*, the SPL query retrieves the maximum *scan_count* value grouped by the user. The following result is produced:

```
Audit.user          max(Audit.scan_count)
------------------------------------------
admin
karun               11475541
n/a                 0
rockyb              109864
splunk-system-user  116088
```

232

CHAPTER 8   LESS-COMMON YET IMPACTFUL SPL COMMANDS

Splitting by _time

You can also split the results by *_time* to create a timechart. You should specify a valid *span* option as well. See the following example:

| tstats avg(server.load_average) FROM datamodel=internal_server **by _time span=1h**

The SPL query utilizes the internal data model *internal_server* to pull the average load. It plots the data points over time with the span of one hour. The resulting line chart is shown in Figure 8-5.

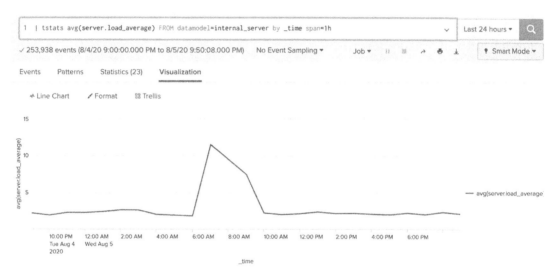

Figure 8-5. *Using tstats to produce a timechart*

If you are planning to use other aggregation commands such as `stats`, `chart`, and `timechart`, you should use the flag *prestats=t*. This option produces results in Splunk internal format that is consumed by the aggregation commands.

233

Caveats with tstats

Wildcards are not allowed in the field names in the aggregate functions. You can use wildcards in key-value pairs to filter data but can't use them in aggregate function. For example, the following SPL will produce an error:

| tstats **avg(*time)** summariesonly=t FROM datamodel=internal_audit_logs BY Audit.user

The SPL query produces the following error:

Error in 'TsidxStats': Wildcards (*) are not supported in aggregate fields

Finally, you cannot use nested eval statements with tstats. For example, the following SPL will produce an error:

| tstats **count(eval(Audit.user="ksmith")** FROM datamodel=internal_audit_logs

Splunk produces the following error:

Error in 'TsidxStats': The tstats / mstats command cannot apply eval function to aggregation function.

If you need to use eval, design your base search in such a way that you can use eval after piping the results of tstats to it. For example:

| tstats **count** FROM datamodel=internal_audit_logs by **Audit.user** | eval myuser = <perform eval checking here> ...

eval coalesce Function

The coalesce function of eval takes in an arbitrary number of arguments and returns the first value that is not null. This can be very helpful if a field is represented using multiple names in different log sources, and you want to normalize the field. For example, consider the following hypothetical events:

… transaction_id = 1234 …
… tran_id = 546 …
… correlationId = 980 …

CHAPTER 8   LESS-COMMON YET IMPACTFUL SPL COMMANDS

By using coalesce function, you can take all the variations of transaction IDs and put them under one common name called *transactionId*. The following eval command will do the job:

```
...| eval transactionId = coalesce(transaction_id,tran_id,correlationId)
```

Splunk assigns the first nonnull value among the three variants to the *transactionId* field. The result is shown in Figure 8-6.

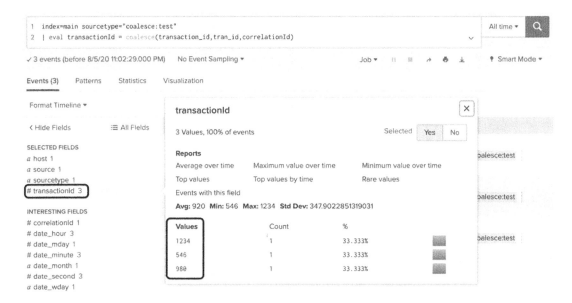

Figure 8-6. *Using eval's coalesce function*

erex

The command erex lets you extract fields from events when you don't know how to form the regular expression. erex command comes up with the regular expression based on the example data you provide. In addition, you can view the regular expression generated by erex so that you can use them in future field extractions. Let us take a look an example. Using Splunk tutorial data, let's say you want to extract the software component from the security events. In the following sample event, *mailsv1* is the software component:

```
Thu Dec 11 2019 00:15:06 mailsv1 sshd[5276]: Failed password for invalid user appserver from 194.8.74.23 port 3351 ssh2
```

CHAPTER 8 LESS-COMMON YET IMPACTFUL SPL COMMANDS

Let's go ahead and use the erex command to extract a new field named *component*. We'll provide *mailsv1* as an example:

```
index=main sourcetype=secure
| erex component examples="mailsv1"
```

You can provide more than one example if available (recommended). Splunk uses the example to form the regular expression and extracts the field *component*. Figure 8-7 shows the extracted field.

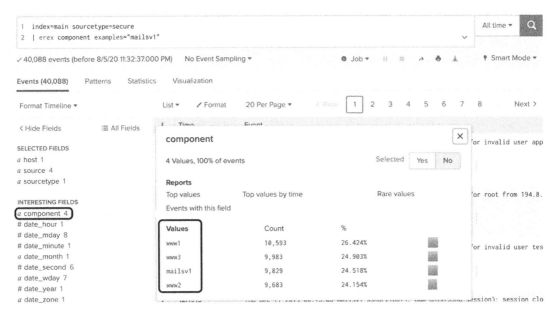

Figure 8-7. *Using erex command to extract fields*

In addition to extracting the field, erex command also reveals the regular expression it came up with under the Job menu. See Figure 8-8.

CHAPTER 8 LESS-COMMON YET IMPACTFUL SPL COMMANDS

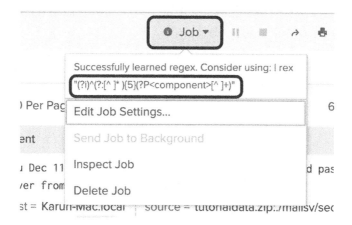

Figure 8-8. *Veiwing the regular expression generated by erex*

You can grab the regular expression and use it in rex command or in field extractions using the *Settings* ➤ *Fields* ➤ *Field extractions* menu.

addtotals and addcoltotals

The command addtotals sums all numerical fields for each search result. For example, using Splunk tutorial data to show the sales of each category, split by the itemId:

```
index=main sourcetype=access_combined_wcookie categoryId != NULL itemId != NULL
| chart sum(sale_price) by categoryId,itemId
| addtotals
```

Figure 8-9 shows the result generated by the chart command followed by the addtotals command.

categoryId	EST-12	EST-13	EST-14	EST-15	EST-16	EST-19	EST-21	EST-27	EST-6	EST-7	OTHER	Total
ACCESSORIES	110.56	138.45	142.43	142.43	111.53	108.56	115.53	133.46	150.39	140.41	423.24	1716.99
ARCADE	1609.32	1759.25	1459.38	1509.36	1924.18	1749.26	1749.25	1599.32	1569.33	1459.38	6462.28	22850.31
SHOOTER	539.73	619.69	519.74	979.51	739.63	659.67	919.54	819.59	779.61	559.72	2458.77	9595.20
SIMULATION	645.62	611.64	696.59	679.60	611.64	679.60	492.71	730.57	475.72	662.61	2106.76	8393.06
SPORTS	441.74	271.84	305.82	237.86	492.71	254.85	424.75	305.82	288.83	271.84	1393.18	4689.24
STRATEGY	2083.84	2335.69	2140.74	1795.83	1854.87	1934.82	1975.93	1522.05	1967.91	2069.84	7198.73	26880.25
TEE	349.50	286.59	293.58	412.41	293.58	370.47	398.43	419.40	328.53	398.43	1377.03	4927.95

Figure 8-9. *Obtaining sum using addtotals*

CHAPTER 8   LESS-COMMON YET IMPACTFUL SPL COMMANDS

The command `addcoltotals` computes the sum of numerical data for previous events and adds a new result row. For example:

```
index=main sourcetype=access_combined_wcookie categoryId != NULL
itemId != NULL
| chart sum(sale_price) by categoryId,itemId
| addtotals
| addcoltotals
```

Splunk produces the result as shown in Figure 8-10.

categoryId	EST-12	EST-13	EST-14	EST-15	EST-16	EST-19	EST-21	EST-27	EST-6	EST-7	OTHER	Total
ACCESSORIES	110.56	138.45	142.43	142.43	111.53	108.56	115.53	133.46	150.39	140.41	423.24	1716.99
ARCADE	1609.32	1759.25	1459.38	1509.36	1924.18	1749.26	1749.25	1599.32	1569.33	1459.38	6462.28	22850.31
SHOOTER	539.73	619.69	519.74	979.51	739.63	659.67	919.54	819.59	779.61	559.72	2458.77	9595.20
SIMULATION	645.62	611.64	696.59	679.60	611.64	679.60	492.71	730.57	475.72	662.61	2106.76	8393.06
SPORTS	441.74	271.84	305.82	237.86	492.71	254.85	424.75	305.82	288.83	271.84	1393.18	4689.24
STRATEGY	2083.84	2335.69	2140.74	1795.83	1854.87	1934.82	1975.93	1522.05	1967.91	2069.84	7198.73	26880.25
TEE	349.50	286.59	293.58	412.41	293.58	370.47	398.43	419.40	328.53	398.43	1377.03	4927.95
	5780.31	6023.15	5558.28	5757.00	6028.14	5757.23	6076.14	5530.21	5560.32	5562.23	21419.99	79053.00

Figure 8-10. Obtaining grand totals using addcoltotals

loadjob

The command `loadjob` lets you load the results from a previously completed search. If the search is a scheduled search, it loads results from the latest run of the scheduled search. This can be especially useful if your search needs to process large amounts of data. With `loadjob`, you are simply retrieving the results of a preexecuted search. So, the results load almost instantly. In addition, you can pipe the results to other commands to further process the data. For example, to load the results of a previously scheduled job

```
| loadjob savedsearch="karun:search:HTTP_Status_Report_By_Action"
```

The SPL query retrieves the results from the last run of the saved search *HTTP_Status_Report_By_Action* which is created by user *karun* in the app *search*. The fully qualified name of the saved search is required, which takes the form of *user:app:saved_search_name*. Splunk produces the result as shown in Figure 8-11.

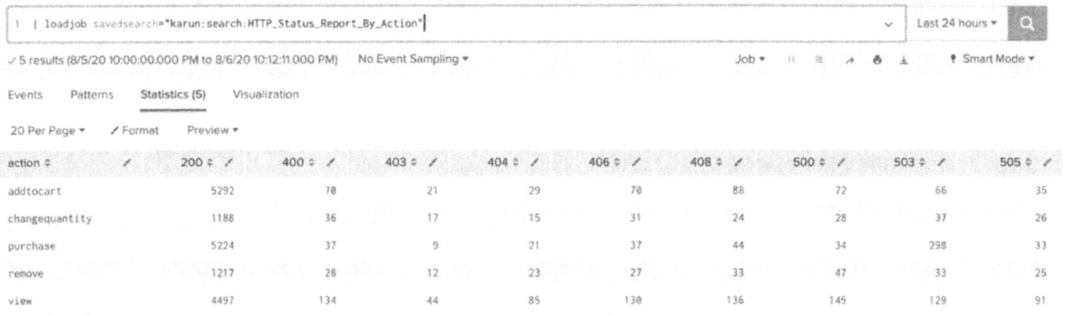

CHAPTER 8   LESS-COMMON YET IMPACTFUL SPL COMMANDS

action	200	400	403	404	406	408	500	503	505
addtocart	5292	70	21	29	70	88	72	66	35
changequantity	1188	36	17	15	31	24	28	37	26
purchase	5224	37	9	21	37	44	34	298	33
remove	1217	28	12	23	27	33	47	33	25
view	4497	134	44	85	130	136	145	129	91

Figure 8-11. Using loadjob to retrieve previously executed scheduled search

You can filter the results of *loadjob* by further processing. For example, to display just the purchase action

```
| loadjob savedsearch="karun:search:HTTP_Status_Report_By_Action"
| search action = "purchase"
```

Note Instead of using the savedsearch name, if you know the search id, you can use the search id with the `loadjob` command.

replace

The `replace` command replaces field values with the value you specify. For example, consider the following SPL query:

```
index=main sourcetype=secure "Failed password"
| top 2 username showperc=f
```

The SPL query retrieves the top two usernames with failed password attempts. The result is shown in the following:

```
Username           count
------------------
Administrator      1020
Db                 965
```

CHAPTER 8   LESS-COMMON YET IMPACTFUL SPL COMMANDS

If you want to change the value *administrator* to *superuser*, you can use the replace command as follows:

```
index=main sourcetype=secure "Failed password"
| top 2 username showperc=f
| replace "administrator" WITH "superuser" IN username
```

The SPL query replaces the value *administrator* with *superuser* in the field *username*. If you omit the field name, all fields are processed. Splunk produces the following result:

```
username        count
---------------
superuser     1020
db             965
```

savedsearch

Using `savedsearch` command, you can kick off a previously saved search. Note that if you specify a time range using the time range picker when using this command, the time range you choose in the time range picker will override the time range saved in the saved search. If you want to use the time range saved in the saved search, you need to select *All time* in the time range picker. For example, to kick off the saved search *HTTP_Status_Report_By_Action*

```
| savedsearch "HTTP_Status_Report_By_Action"
```

Unlike `loadjob` command which retrieves the results from the last run of the previously executed job, `savedsearch` command runs the specified search and generates the results. In addition, you don't need to specify the fully qualified job name such as *user:app:saved_search_name or search id*. You just need to specify the saved search name. Splunk produces the result as shown in Figure 8-12.

CHAPTER 8 LESS-COMMON YET IMPACTFUL SPL COMMANDS

action	200	400	403	404	406	408	500	503	505
addtocart	5292	70	21	29	70	88	72	66	35
changequantity	1188	36	17	15	31	24	28	37	26
purchase	5224	37	9	21	37	44	34	298	33
remove	1217	28	12	23	27	33	47	33	25
view	4497	134	44	85	130	136	145	129	91

Figure 8-12. Using savedsearch command to run a saved search

As we wrap up this chapter, I would like to point out that there are many other commands in the SPL repertoire that you may find useful. I left out commands that are only used in specialized situations. For example, I did not discuss metrics store-related commands such as mstats and mcollect. I encourage you to review Splunk product documentation if you need to learn more about them. Now, let's review the key takeaways from this chapter.

Key Takeaways

1. The commands iplocation and geostats help with plotting data in a geographic map.

2. Use untable command to create stats-style tabular output and xyseries command to perform the reverse.

3. The bin command can be used to put numerical data into discrete sets.

4. In order to generate statistics from accelerated data models and indexed fields, use tstats command.

5. With tstats, you cannot use wildcards in the field names used for aggregation functions.

6. Use the datamodel command to view the hierarchy of the dataset in a datamodel.

7. In order to sum the values in each result, use `addtotals`. The command `addcoltotals` sums the values in each column and adds a new row at the end of the results.

8. You can use `erex` command to have Splunk generate the regular expressions that can be used for field extraction.

9. To load the results from the last run of a search, use `loadjob` command.

10. The command `savedsearch` can be used to kick off a saved search.

In the next chapter, we'll discuss how to optimize your SPL queries for performance.

CHAPTER 9

Optimizing SPL

We have come to the final chapter of this book. So far, we have learned how to use the many useful SPL commands. In this chapter, last but not least, we are going to learn how to optimize your SPL queries for maximum performance. If you have been using Splunk for a while, you would have inevitably come across situations where your search seemed to be taking forever to complete. There are many actions you can take to speed up your search. Tuning your search is an important and very useful step for several reasons. First, it gets you the result you want faster. Second, it can prevent miscalculations due to time-out of long running searches. For example, by default, a subsearch will time out at 60 seconds. It also decreases the processing power required by the search head and/or search peers. In this chapter, we'll take a thorough look at how to measure the effectiveness of your search and review tools that aid you in troubleshooting search performance.

Tip You do not need to optimize all searches you write. While optimizing all of your searches may be beneficial, I recommend you only focus on the frequently used searches (such as a scheduled search) or searches that retrieve large amounts of data. These are the searches that will provide the greatest value for your time invested in tuning them.

Factors Affecting Performance of SPL

While building searches, we are often focused on the results alone, without considering the performance impact. This may not seem like a big problem at first. However, if you end up scheduling the search to be run frequently in an automatic fashion, the performance of the search becomes more important. In addition, if the underlying amount of data increases, the time taken by the search will usually increase as well. The first step in optimizing SPL is to fully understand the factors that affect the performance. Let's take a look at them.

CHAPTER 9   OPTIMIZING SPL

Quantity of Data Moved

The amount of data moved directly impacts the search performance. Regardless how well a SPL query is tuned, if the amount of data that Splunk has to read from the disk is large, the time taken for the search to complete will also be proportionally high. As you know, data is stored in indexes. Indexes are managed by a set of Splunk servers called *indexers*. The job of indexers in Splunk is twofold. First, it receives data from the machine data source and transforms them into searchable events (this process is known as indexing). Second, it searches the indexes upon receiving a search request. In a distributed Splunk environment (most production Splunk environments are distributed), you will have a cluster of indexers with multiple copies of data stored among them. A set of Splunk servers called *search heads* provide the user interface for the users to run searches. Search heads distribute the searches to indexers (also known as search peers). Indexers run the searches and return results to the search head. Search head merges the results and presents it to the end user. As you can tell, the amount of data read by the indexers will directly impact the search time. Horizontally scaling the Splunk indexer cluster is one way to handle the problem with moving large amounts of data. When horizontally scaled, many indexers can operate in parallel to share the search load.

Time Range of the Search

Time range of your search is one of the biggest factors that will affect the performance. The wider the time range, the larger the time taken to run the search. This is because generally wider time ranges tend to move more data as it needs to cover longer time frames. In addition, Splunk organizes data in data buckets that are time oriented. The newest data is stored in *hot* buckets. Once the data in hot buckets age out, they are moved to *warm* buckets. From there, it is moved to *cold* buckets as they age out further. Finally, the data is moved to *frozen* state where it is either deleted or archived. In many large Splunk environments, the hot and warm buckets are hosted on faster storage, but cold and frozen data are stored in slower and cheaper storage. This is to take advantage of the fact that older data is less frequently searched than newer data. The age-out process is completely customizable by the Splunk administrators.

Splunk Server Resources

Splunk indexers are very i/o intensive. That is because they generally have to sift through lots of data during searches. In addition, the indexers constantly write new events to disk. The IOPS (input/output operations per second) of the disk subsystem of Splunk servers play a critical role in search performance. As per Splunk's reference hardware, the minimum IOPS required for a production Splunk environment is 1200. However, in large environments, it can be a lot higher than that. In addition, as mentioned previously, horizontally scaling Splunk indexers help with distributing the search and indexing load and thereby increasing search performance.

Unfortunately, if you don't have direct responsibility of owning the Splunk server infrastructure, you may not have control over Splunk server resources. But if you observe severe search performance degradation, consult with your Splunk administrators.

Optimizing Searches

There are many steps you can take to optimize your steps. Let's review them now.

Use Fast or Smart Search Mode

When performing searches, pay attention to the search mode. When it comes to time taken to execute the search, fast mode yields the best performance. Verbose mode yields the worst. And smart mode provides the blended approach. Figure 9-1 shows how to select the search modes.

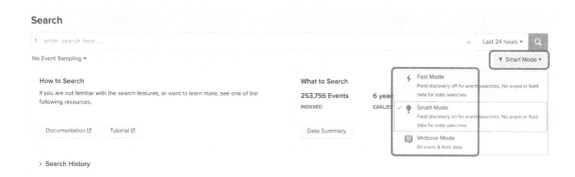

Figure 9-1. Selecting the search mode

CHAPTER 9 OPTIMIZING SPL

With fast mode, the field discovery is turned off. But it provides the fastest performance. Note that you can still use the fields in your searches if you know they are going to be there in the search results. With verbose mode, field discovery is turned on. In addition, verbose mode retrieves all events and makes it available in Events tab during statistical searches (such as `stats`). With smart mode, field discovery is turned on, but events data is not retrieved during statistical searches.

Note that the search mode is preserved when a search is saved as a report. If you have a slowly performing scheduled search, for example, it is useful to review the search mode of the searches powering the search. Update them to fast or smart mode for better performance.

Narrow Down Time Range

Whenever possible, use as narrow as a time frame as possible. As shown in Figure 9-2, in the time picker, you can choose the data and time range to provide as narrow as a time frame as possible.

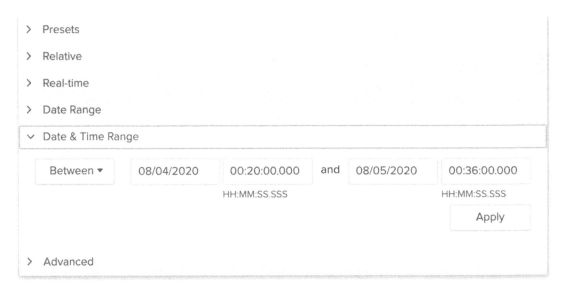

Figure 9-2. Using the Date & Time Range of time picker

Filter Data Before the First Pipe

The fragment of the SPL code before the first pipe is called base search. Include as much filtering as possible in the base search. When a search is executed, Splunk uses the terms in the base search to create a bloom filter, which is a binary array generated off of a hashing algorithm. This bloom filter is compared against bloom filters of data buckets. Splunk opens these buckets only when there is a match found. This means by providing as much filtering as possible before the first pipe, you effectively minimize the number of data buckets opened. Consider the following example:

```
index="main" sourcetype="access_combined_wcookie" error
| timechart span=1h count
```

By filtering for the string error, the base search can potentially reduce the number of buckets to open, and number of events to retrieve.

Use Distributable Streaming Commands Ahead in the Pipeline

Most Splunk environments employ a distributable environment in which one or more Splunk servers designated as search heads provide the user interface for running searches. As previously stated, search heads dispatch searches to a set of back-end servers known as Splunk indexers. It is the indexers that execute the search and retrieve the events from the indexes. Figure 9-3 depicts this architecture.

CHAPTER 9   OPTIMIZING SPL

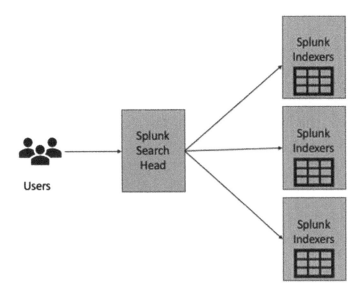

Figure 9-3. A basic Splunk distributed environment

When it comes to SPL performance, there are three major types of commands to consider. *Transforming* commands are those that operate on the entire data set. They are always executed on the Splunk search head. Examples include `stats`, `timechart`, `top`, and `rare`. *Centralized* streaming commands do not require the entire data set, but the order of the events is important. They are always run on the search head. Examples include `streamstats`, `transaction`, and `head`. *Distributable* streaming commands are those that do not require the entire result set (like centralized streaming commands), but event order is not important (unlike centralized streaming commands). They can be run on the indexers if all preceding search commands are executed on the indexers. If any of the preceding search commands are executed on the search head, the distributable streaming command is also executed on the search head. Examples include `eval`, `rename`, `fields`, `regex`, and `replace`. Consider the following example:

```
index="main" sourcetype="secure"
| stats count by source,username
| rename username AS customerName
```

In the SPL query, `stats` is a transforming command which is always executed on the Splunk search head. The command `rename` is a distributable streaming command that can be run on the indexers. However, since a preceding search command (`stats`)

248

is run on the search head, `rename` is also run on the search head. In order to get a better performance, rearrange the SPL as follows:

```
index="main" sourcetype="secure"
| rename username AS customerName
| stats count by source,customerName
```

Because `rename` is a distributable streaming command and there is no previous search command that runs on the search head, `rename` is executed on the indexers as the events are retrieved from the index.

Best Practices for Scheduling Searches

As you work with Splunk, you will find yourself scheduling searches quite often. Scheduling the searches enables you to periodically run the searches automatically without you having to run them manually. Further, you can configure scheduled searches to send the results via email or even post the results to a remote website. Search performance becomes very critical when you schedule your searches, especially if you are running them very frequently. Due to poor performance, searches may get skipped if the previous instance of that search is still running. I've listed some of the best practices in the following to handle scheduled searches.

Stagger Your Searches

If you have many scheduled searches, do not start all of them at the same time. Consider staggering them so that they all don't end up kicking off at the same time. For example, instead of starting all your daily jobs to start at 6:00 AM, you could stagger them by scheduling to start at 6:00, 6:25, 6:50, and so on.

Your Splunk administrator may have some limits on the number of simultaneous searches you can run (typically 3-10). If you exceed the limit, your search may get queued. Staggering your searches decreases the chance that you will be in a situation where you are running many searches at the same time.

CHAPTER 9   OPTIMIZING SPL

Use cron for Maximum Flexibility

When scheduling searches in Splunk Web, you have to pick a schedule for the search to be executed. You can either choose from a list of preset options or choose cron to create your schedule using a cron expression. I recommend using a cron expression as it provides the maximum flexibility to specify time. For example, you cannot schedule with minute granularity using the preset option. As shown in Figure 9-4, your only options are 0, 15, 30, and 45 minutes past the hour.

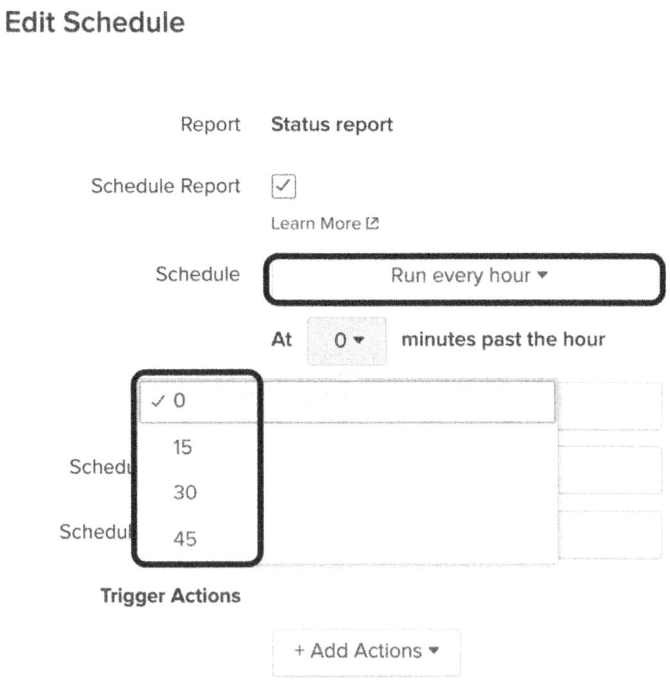

Figure 9-4.   Using preset time does not provide a lot of flexibility when scheduling a search

However, when you choose *Run on Cron Schedule* from the drop-down, you can specify a cron expression for the schedule. See Figure 9-5.

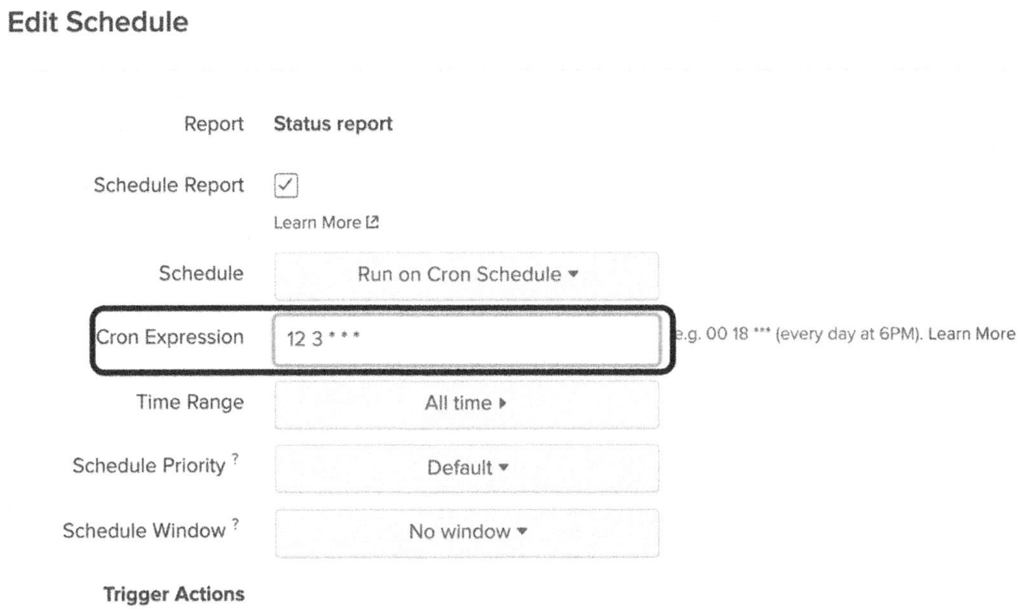

Edit Schedule

Report	**Status report**
Schedule Report	☑ Learn More
Schedule	Run on Cron Schedule ▼
Cron Expression	12 3 * * * e.g. 00 18 *** (every day at 6PM). Learn More
Time Range	All time ▶
Schedule Priority ?	Default ▼
Schedule Window ?	No window ▼

Trigger Actions

Figure 9-5. *Using cron expression when scheduling a search*

For example, the cron expression 12 3 * * * indicates the schedule of 3:12 AM every day. From left to right, description of the five fields of the cron expression is provided here:

- Minute (0-59).
- Hour (0-23).
- Day of the month (1-31).
- Month (1-12).
- Day of the week (0-6); 0 is Sunday.

In addition, the following special formats are also used:

- */N means every N value in this field. For example, */5 in the minute field indicates every 5 minutes starting from 0.

- N,M means comma-separated list of values. For example, 0, 6, 12, and 18 in the hour field means 12 AM, 6 AM, 12 PM, and 6 PM.

- I-J means a value range including the range start and end. For example, 6-8 in the hour field means 6 O'clock, 7 O'clock, and 8 O'clock.

CHAPTER 9   OPTIMIZING SPL

Utilize Schedule Window Setting

You can utilize the automatic schedule window feature of Splunk when scheduling searches. The schedule window lets Splunk choose the best time to run your searches once it is time to run the search. See Figure 9-6.

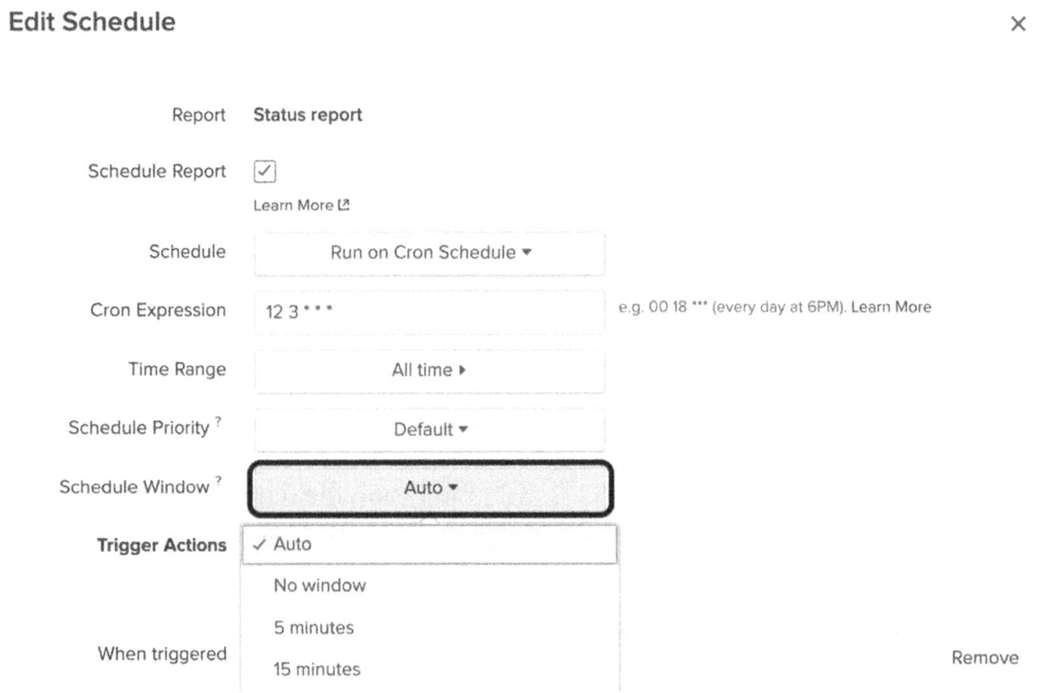

Figure 9-6. Choosing automatic schedule window for scheduled searches

You can create certain Splunk knowledge objects to speed up your search. Let's review them next.

Useful Splunk Knowledge Objects to Speed Up Searches

Splunk knowledge objects are entities that you can create to enrich and manage your machine data. Some of the most common examples of knowledge objects include reports, dashboards, alerts, lookups, and field extractions. There are three knowledge objects I will describe here that can be used to speed up searches. They are accelerated reports, summary indexes, and accelerated data models.

Accelerated Reports

When you save a search, which is also known as a report, you have the option to accelerate it. When you accelerate a report, Splunk periodically runs the report in the background and builds data summaries. Splunk stores the summaries on the indexers, alongside the index bucket directories. Next time when you run the report, instead of retrieving raw data from the index, results are retrieved from the data summaries. Because the summary is smaller than the index, as it only contains precomputed statistics, your searches run faster. In order to accelerate a report, it must use transforming commands such as stats, timechart, and chart. And the search mode must be either fast or smart. In addition, your Splunk administrator must have granted necessary permissions to configure report acceleration. To accelerate a report, simply choose *Edit* ➤ *Edit acceleration* in Splunk Web. When enabling report acceleration, you should also select the summary range which sets the range of time for which the report is accelerated. Figure 9-7 shows how you can accelerate a report.

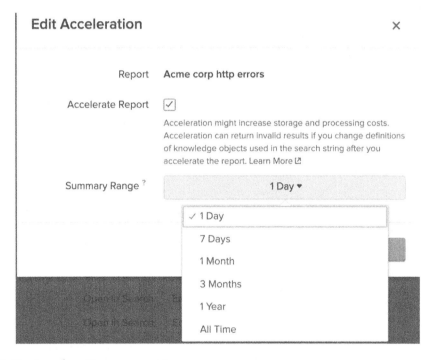

Figure 9-7. Accelerating a report

CHAPTER 9   OPTIMIZING SPL

> **Note** Accelerated reports cannot be edited. If you need to edit the report, you must remove the acceleration first.

Next, let's take a look at summary indexes.

Summary Indexes

When you are unable to use report acceleration, summary indexes can be an alternative. With summary indexing, you schedule your searches and configure them to send the results to a summary index. You decided what exactly goes into the summary index. Summary indexes are stored on the indexers just like any other indexes. Future searches must be run against the summary indexes instead of the original index. Because summary indexes are much smaller than original indexes as they just store the summarized statistics, they run much faster. You can enable summary indexing by navigating to *Settings* ➤ *Searches, reports and alerts* and selecting *Edit summary indexing* for your report. Figure 9-8 shows how to enable summary indexing.

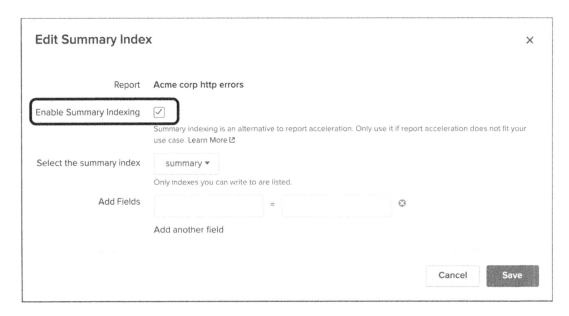

Figure 9-8. Enabling summary indexing

There is a caveat with summary indexes though. If the scheduled search that populates the summary index does not run for some reason, the summary index will have gaps. The summary index must be manually backfilled by rerunning the scheduled search using *fill_summary_index.py* script. To invoke the script, use the following command:

```
$SPLUNK_HOME/bin/splunk cmd python fill_summary_index.py
```

Note Data in summary indexes are stored with the default sourcetype *stash*. There is no additional Splunk license cost when storing data in summary indexes with this sourcetype.

In addition to using a scheduled search to populate summary indexes, you can also use the command `collect` to send results of a search to a summary index. Note that if the sourcetype is anything other than *stash* in the data stored in the summary index, the data is counted toward license usage.

We'll now take a look at the third and preferred option for speeding up searches – accelerated data models.

Accelerated Data Models

A data model is a hierarchically structured data set that you generate using search constraints you design. You can create data models from *Settings ➤ Data models* menu. When you accelerate a data model, Splunk builds summaries in the form of TSIDX files (time-series index files) on the indexers. Searches against accelerated data models must be run using `tstats` command. Because the search is run against the preorganized accelerated data models, they run much quicker compared to running against the original index. Splunk automatically updates TSIDX files every 5 minutes and removes outdated ones every 30 minutes. When a data model is accelerated, all fields in the data model become indexed fields that are available for searching. You can use the SPL command `datamodel` to review the structure of a data model.

A major advantage of accelerated data models over summary indexing is that there is no need of manual upkeep. Splunk automatically ensures that there are no gaps in the summaries. Compared to report acceleration, data model provides a lot of flexibility as each field in the data model becomes available for faster searching. In report acceleration, you simply speed up a predefined report.

CHAPTER 9   OPTIMIZING SPL

Note   Only Splunk administrators can enable acceleration on a data model. And once a data model is accelerated, you cannot edit the data model.

To accelerate a data model, first open the data model by navigating to *Settings* ➤ *Datamodels*. Then, select *Edit* ➤ *Edit Acceleration*. When enabling data model acceleration, you should also need to select the summary range which sets the range of time for which the data model is accelerated. Figure 9-9 shows how to accelerate a data model.

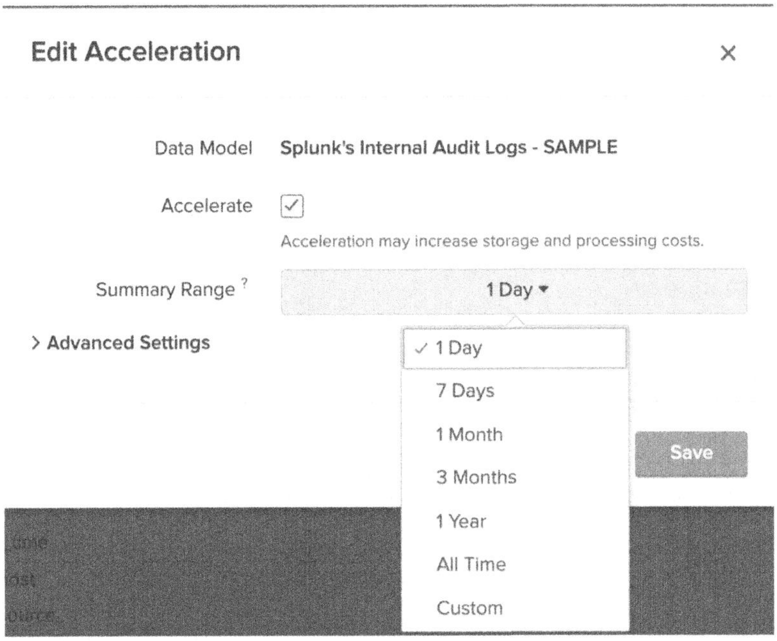

Figure 9-9. Accelerating data models

Among the three options I presented, report acceleration, summary indexing, and data model acceleration, I recommend using the data model acceleration. Next, we'll take a look at a useful tool that Splunk provides to examine the execution of a search – job inspector.

CHAPTER 9   OPTIMIZING SPL

Using Job Inspector

Using job inspector, you can examine the inner workings of a search. It is an excellent troubleshooting tool to diagnose poorly performing searches. You can compare job inspector to a relational database platform's SQL explain plan, which shows the execution costs of various fragments of the query. You can invoke the job inspector either while the search is running or after it is completed. Click the *Job* ➤ *Inspect Job* to bring up the job inspector. See Figure 9-10.

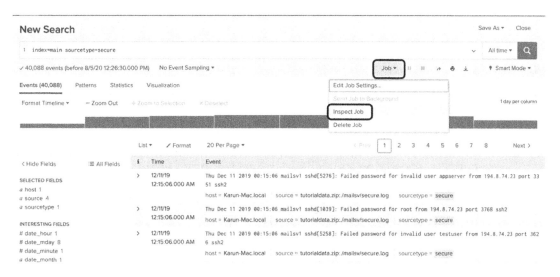

Figure 9-10. *Invoking job inspector*

The job inspector opens in its own pop-up window. As seen on Figure 9-11, at the top it provides basic run time information about the search, such as the total time taken and number of results returned. In addition, it has two sections – execution costs and search job properties.

257

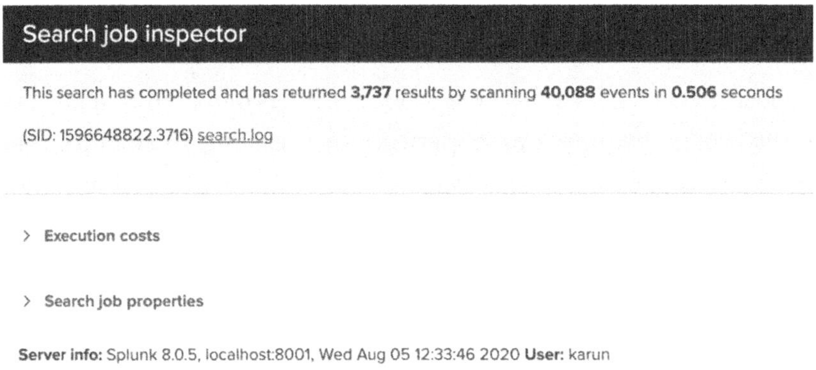

CHAPTER 9   OPTIMIZING SPL

> **Search job inspector**
>
> This search has completed and has returned **3,737** results by scanning **40,088** events in **0.506** seconds
>
> (SID: 1596648822.3716) search.log
>
> \> Execution costs
>
> \> Search job properties
>
> **Server info:** Splunk 8.0.5, localhost:8001, Wed Aug 05 12:33:46 2020 **User:** karun

Figure 9-11. Search job inspector

The job inspector generally contains lots of data that may not be readily understandable. You don't have to analyze every piece of information in the job inspector.

The execution costs section shows how long each component of the SPL query took to complete. The most important ones I recommend are provided in the following table.

Table 9-1. Important sections of execution costs in job inspector

Execution cost	Description
command.search.index	Time taken to examine the index files (TSIDX files) to determine the events to be retrieved
command.search.rawdata	Time taken to retrieve the actual events from raw data
command.search.kv	Time taken to apply field extractions
command.search.lookups	Time taken to perform lookups
command.search.typer	Time taken to apply event types

In addition, each command used in the SPL will have a specific execution cost in the format *command.<command name>*. This can help with identifying any particular command that is taking too long to execute. Figure 9-12 shows the execution costs of a sample search.

CHAPTER 9 OPTIMIZING SPL

Search job inspector

This search has completed and has returned **3,737** results by scanning **40,088** events in **0.607** seconds

(SID: 1596649958.3722) search.log

Execution costs

Duration (seconds)	Component	Invocations	Input count	Output count
0.00	command.fields	19	40,088	40,088
0.43	command.search	19	-	40,088
0.03	command.search.expand_search	2	-	-
0.00	command.search.calcfields	16	40,088	40,088
0.00	command.search.expand_search.calcfield	2	-	-
0.00	command.search.expand_search.fieldaliaser	2	-	-
0.00	command.search.expand_search.kv	2	-	-
0.00	command.search.expand_search.lookup	2	-	-
0.01	command.search.expand_search.sourcetype	2	-	-
0.02	command.search.filter	16	-	-
0.02	command.search.index	20	-	-
0.00	command.search.fieldalias	16	40,088	40,088
0.00	command.search.index.usec_1_8	178	-	-
0.00	command.search.index.usec_8_64	17	-	-
0.12	command.search.rawdata	16	-	-
0.12	command.search.typer	16	40,088	40,088
0.11	command.search.kv	16	-	-
0.01	command.search.lookups	16	40,088	40,088

Figure 9-12. *Viewing the execution costs of a job in the job inspector*

For example, the *command.search.kv*, which denotes the time taken by field extractions, took 0.11 seconds. When you are troubleshooting search performance, you may want to look into field extractions and automatic lookups that may be adding time to the overall search time.

The search job parameters provide information about the search job. The most important parameters are shown in Table 9-2.

259

Table 9-2. Important sections of search job parameters in job inspector

Search job parameter	Description
optimizedSearch	The restructured search based on built-in search optimizer. Splunk automatically moves commands around to make the search faster
scanCount	Number of events read
resultCount	Number of results returned

Tip To measure the performance of a search, I recommend using *scanCount* instead of *resultCount*. Your goal should be to limit the number of events scanned to improve the performance.

Now, let's review the key takeaways from this chapter.

Key Takeaways

1. The key to fast searches is moving as little data as possible.
2. Time range is the most powerful lever you have to control the amount of data being moved.
3. Not all searches need to be tuned. Focus on frequently used searches and searches that retrieve large amounts of data.
4. Use the search modes in the following order – fast, smart, and verbose – to meet your needs. When you save a search, the search mode is preserved.
5. Filter as much data as possible before the first pipe in the SPL query.
6. When scheduling searches, make use of the automatic schedule window.
7. Utilize report acceleration, data model acceleration, and summary indexing for searching against large amounts of data.

8. Data model acceleration is more efficient than report acceleration and summary indexing.

9. Use job inspector to analyze the execution costs of various parts of your SPL query.

10. Rearrange the components of the SPL query so that distributable streaming commands are placed before any transforming commands.

This concludes our journey together in learning the practical use of Splunk Search Processing Language. We have covered a lot of material, and it is quite possible you are a bit overwhelmed. The key to mastering SPL is to practice as much as you can. Over time, many of the commands we discussed will become second nature to you. You can always refer to Splunk documentation at docs.splunk.com for up-to-date information on SPL.

Happy Splunking!

Index

A
addtotals/addcoltotals command, 237, 238
Appendcols command, 135, 136
Append command, 132–135
Appendpipe command, 136–138
Artificial Intelligence for IT Operations (AIOps), 7

B
bin command, 228

C, D
cluster command, 196–198
coalesce function, 234, 235
convert command, 202–204
Correlating and grouping data, 113

E
erex command, 235–237
Extracting fields
    accessing process, 161
    extractor wizard
        events, 154
        event selection, 155
        fields screen, 156, 157
        permissions, 158
        preview screen, 157
        regular expression, 156
        results, 154, 155
        search interface, 159
        success message screen, 159
        validate screen, 157, 158
    inline, 163
    iplocation command, 164
    literal/metacharacters, 165
    meaning, 153
    navigation menu, 160
    regular expression, 163, 165–167
    rex command, 168–171
    sidebar, 164
    structured data
        JSON data, 215
        rendering data, 215
        spath command, 216–219
        XML format, 214
    transform, 163

F
Factors affecting performance, 243
    data moved/quantity, 244
    indexers, 244
    server resources, 245
    time range, 244
Fields
    addtocart/purchase, 144
    default fields, 150, 151
    discovery, 152, 153

INDEX

Fields (*cont.*)
    duplicates, 174, 175
    extraction (*see* Extracting fields)
    filtering, 171, 172
    flexible schema, 146–148
    index-time *vs.* search-time, 148, 149
    informative charts, 144–146
    internal fields, 150, 151
    literal string, 142
    regular expression, 147
    schema-on-read, 146
    schema-on-write, 146
    smart/verbose mode, 152
    sorting, 173
    tailored searches, 141–144
File-based lookup, 179
fillnull/filldown command, 200, 201

G

geostats command, 225, 226
Group and correlate data, *see* Transaction command

H

Handling multi-valued fields
    makemv, 204–207
    mvcombine, 209, 210
    mvcount, 210
    mvexpand, 208
    mvfilter, 211, 212
    mvfind, 212
    mvindex, 211
    mvjoin, 213
    mvsort() eval function, 213
    nomv, 207, 208
    split function, 214

I

Index-time *vs.* search-time fields, 148, 149
Internal fields, 150, 151
Interquartile range (IQR), 199
iplocation command, 164, 223–225

J

Job inspector
    execution costs, 257, 258
    parameters, 260
    search job inspector, 257, 258
    viewing option, 259

K

kmeans command, 195, 196
Knowledge objects, *see* Speed up searches

L

loadjob command, 238
Lookups
    automatic lookup, 189, 190
    command, 184–187
    external data, 177
    file-based lookup, 179
    join/match field, 179
    maintaining data, 186
    messages, 188
    outputlookup command, 187
    OUTPUT option, 185
    table creation
        content verification, 182, 183
        definition, 181
        file information, 180

INDEX

file link, 179
inputlookup command, 182
providing details, 181
types, 178

M, N

Machine data, 1
AIOps technologies, 7
business analytics, 6
categories of, 2
distribution, 8
events, 3
IT operations and
monitoring, 5, 6
logs, 3
metrics, 5
schema-on-read technology, 8
security/SIEM, 6
size, 7
speed, 7
structure, 8
time-series data, 5
traces, 4
value, 5

O

Operational data intelligence
add Knowledge function, 9
collect and Index function, 9
functions, 8–10
high-level architecture
components, 11, 12
heavy forwarder, 12
indexer, 11
search head, 11
monitor and alert function, 10

report and visualize
function, 10
search and investigate
function, 9
transforming commands, 10
Outlying values, 199–201

P, Q

predict command, 193–195

R

Transaction command
append, 132–135
appendcols, 135, 136
appendpipe, 136–138
constraints, 120
maxevents, 121
maxpause, 121
maxspan, 120
field values, 114–117
find unfinished
tractions, 124–126
grouping events, 114
itemId field, 122
joins
limitations, 132
sourcetype, 130, 131
multiple events, 122
multivalued field, 123
output of, 115
strings, 117–120
subsearches, 126
construction, 126–128
pitfalls, 128
splunkd process, 127
replace command, 239, 240

265

INDEX

S

savedsearch command, 240, 241
Scheduling searches
    cron expression, 251, 252
    schedule window, 252
    staggering, 249
Search optimization
    distributable streaming commands, 247–249
    fast/smart search modes, 245, 246
    filtering data, 247
    narrow down time range, 246
Search peer, 11, 179, 243, 244
Search Processing Language (SPL), 1
    arithmetic operators, 17
    assistant mode, 28–30
    boolean operators, 16
    commands, 13
    comparison operators, 15
    event details, 35–37
    fields sidebar, 34, 35
    functions, 16
    key-value pairs, 14
    literal strings, 13
    logical operators, 15
    meaning, 12
    operational data intelligence, 9
    preferences menu, 29
    query, 23
    resulting screen, 32, 33
    search mode
        fast mode, 30
        selection process, 30
        smart mode, 31
        verbose mode, 31
    search pipeline, 18
    security/SIEM, 6
    syntax, 12
    timeline, 33, 34
    time picker screen, 31, 32
    tutorial data, 23
        adding data, 24
        download zip file, 23
        indexing process, 27
        input settings screen, 26
        launching screen, 28
        review screen, 27
        tutorialdata.zip selection, 25
        uploading data, 24
    wildcard asterisk (*), 14
Software as a Service (SaaS), 19
Speed up searches
    accelerated reports, 253
    data model, 255, 256
    indexes, 255, 256
Statistics (Stats)
    aggregate functions, 47, 48
    chart, 58–62
    counting events
        command process, 42
        distinct_count function, 44
        eval expressions, 43, 44
        field, 43
    eval command
        calculate values, 63
        conditional operations, 67–69
        converting (numbers/strings), 64
        data information, 62
        expression, 63
        formatting, 66, 67
        rounding numbers, 67
    event order functions, 52
    events, 40
    eventstats command, 52, 53
    spitting results, 45–47

INDEX

streamstats command, 54–56
syntax, 40–42
time-based functions, 50, 51
top and rare commands, 56–58
transforming command, 39
unique value, 48–50
visualizations (*see* Visualizations)

T

Time-series database, 81
- arithmetic operations, 109, 110
- current-day metrics, 106–109
- data buckets, 81
- date_time fields, 97–99
- modifiers
  - snap-to time unit, 100–102
  - time-picker menu, 99
  - units and abbreviations, 100
- retrieving logs
  - browser window, 95, 96
  - nearby events selection, 96
  - proximity, 95
  - search results, 95
- search interface, 82
- timechart
  - aggregation functions, 87–89
  - column chart visualization, 92
  - HTTP errors and visualization, 91
  - identifiers, 87
  - line chart, 85
  - nullstr, 91
  - plot relevent series, 94
  - span specification, 85–87
  - split-by field, 89–91
  - stack mode selection, 93
  - status codes, 93
- tutorial data, 84
- visualization tab, 84
- time periods
  - multiple time series, 104
  - series option, 104, 105
  - timechart, 103
  - timewrap command, 103
  - wrapping span, 104
- zone configuration, 83

tstats command
- aggregate functions, 234
- meaning, 230
- query accelerated data models, 231
- splitting by _time_, 233
- where and by clause, 230

U

untable command, 226
User interface, 18
- installation, 19
- search/reporting app, 18
- web launcher app
  - app bar, 22
  - home page, 19
  - icon and label app, 22
  - search bar, 22
  - search history, 22
  - search interface, 20
  - Splunk bar, 21
  - time range picker, 22

V, W

Visualizations
- meaning, 69
- SPL commands, 71

267

INDEX

Visualizations (*cont.*)
   statistics tab, 70
   switching options
      area chart, 73, 74
      bar chart, 72, 74, 75
      column chart, 74, 76, 77
      line chart, 72, 73
      pie chart, 75
      plotting multiple data series, 76–78
      timechart command, 71
      visualizations tab, 70, 71

X, Y, Z

xyseries command, 227

Made in the USA
Monee, IL
15 April 2021